Humanities for the Environment

Humanities for the Environment, or HfE, is an ambitious project that from 2013–2015 was funded by a generous grant from the Andrew W. Mellon Foundation. The project networked universities and researchers internationally through a system of 'observatories'. This book collects the work of contributors networked through the North American, Asia–Pacific and Australia–Pacific observatories.

Humanities for the Environment showcases how humanists are working to 'integrate knowledges' from diverse cultures and ontologies and pilot new 'constellations of practice' that are moving beyond traditional contemplative or reflective outcomes (the book, the essay) towards solutions to the greatest social and environmental challenges of our time. With the still controversial concept of the 'Anthropocene' as a starting point for a widening conversation, contributors range across the geographies, ecosystems, climates and weather regimes; moving from icy, melting Arctic landscapes to the bleaching Australian Great Barrier Reef, and from an urban pedagogical 'laboratory' in Phoenix, Arizona, to Vatican City in Rome. Chapters explore the ways in which humanists, in collaboration with communities and disciplines across academia, are responding to warming oceans, disappearing islands, collapsing fisheries, evaporating reservoirs of water, exploding bushfires and spreading radioactive contamination.

This interdisciplinary work will be of great interest to scholars in the humanities, social sciences and sciences interested in interdisciplinary questions of environment and culture.

Joni Adamson is Professor of Environmental Humanities in the Department of English, and Director of the Environmental Humanities Initiative at the Julie Ann Wrigley Global Institute of Sustainability at Arizona State University, USA.

Michael Davis is an Honorary Research Fellow with the Sydney Environment Institute, University of Sydney, Australia, and a Member of the Australian Institute of Aboriginal and Torres Strait Islander Studies.

Routledge Environmental Humanities

Series editors: Iain McCalman and Libby Robin

The *Routledge Environmental Humanities* series is an original and inspiring venture recognising that today's world agricultural and water crises, ocean pollution and resource depletion, global warming from greenhouse gases, urban sprawl, overpopulation, food insecurity and environmental justice are all *crises of culture*.

The reality of understanding and finding adaptive solutions to our present and future environmental challenges has shifted the epicentre of environmental studies away from an exclusively scientific and technological framework to one that depends on the human-focused disciplines and ideas of the humanities and allied social sciences.

We thus welcome book proposals from all humanities and social sciences disciplines for an inclusive and interdisciplinary series. We favour manuscripts aimed at an international readership and written in a lively and accessible style. The readership comprises scholars and students from the humanities and social sciences and thoughtful readers concerned about the human dimensions of environmental change.

Humanities for the Environment

Integrating knowledge, forging new constellations of practice

Edited by
Joni Adamson and Michael Davis

Routledge
Taylor & Francis Group
LONDON AND NEW YORK

earthscan
from Routledge

First published 2017
by Routledge
2 Park Square, Milton Park, Abingdon, Oxon OX14 4RN

and by Routledge
711 Third Avenue, New York, NY 10017

Routledge is an imprint of the Taylor & Francis Group, an informa business

British Library Cataloguing-in-Publication Data
A catalogue record for this book is available from the British Library

Library of Congress Cataloging-in-Publication Data
Names: Adamson, Joni, 1958- editor. | Davis, Michael (Michael Barry) editor.
Title: Humanities for the environment : integrating knowledge, forging new constellations of practice / edited by Joni Adamson and Michael Davis.
Description: New York, NY : Routledge, 2016. | Includes bibliographical references.
Identifiers: LCCN 2016025436| ISBN 9781138188167 (hb) | ISBN 9781315642659 (ebook)
Subjects: LCSH: Human ecology--Case studies. | Environmental sciences--Social aspects--Case studies. | Science and the humanities--Case studies. | Nature--Effect of human beings on--Case studies. | Climatic changes--Social aspects--Case studies.
Classification: LCC GF51 .H77 2016 | DDC 304.2--dc23
LC record available at https://lccn.loc.gov/2016025436

ISBN: 978-1-138-18816-7 (hbk)
ISBN: 978-1-315-64265-9 (ebk)

Typeset in Goudy
by Saxon Graphics Ltd, Derby

Contents

Figures

Acknowledgements

We would like to thank Iain McCalman and Libby Robin, editors of the Routledge Environmental Humanities Series, for their own significant contributions to the fast emerging field of the environmental humanities, and for their strong support of this project.

We thank the institutions that afforded us the time, support and funding to incubate our ideas within the larger, international research space of global observatories organised by the Humanities for the Environment (H*f*E) project. H*f*E meetings held in Arizona, Australia, Hong Kong and Taiwan were generously funded by the Andrew W. Mellon Foundation, the Consortium for Humanities Centers and Institutes, the Office of the President at Arizona State University, the Institute for Humanities Research at Arizona State University, the Sydney Environmental Institute at the University of Sydney, the Center for Humanities Innovation and Social Practices at National Sun Yat-sen University and the Humanities Center at National Chung Hsing University. We are grateful for the opportunities this support offered some of our contributors to deliver keynotes and presentations that are expanded for this volume, or to work together on collaborative projects described in our chapters.

As documented in our introductory chapters (1 and 4), the Environmental Humanities emerged out of nearly two decades of organising, predating the 2013 launch of the Humanities for the Environment project. We will not repeat that history here, but instead, thank the professional organisations that have offered our contributors vibrant scholarly homes in which to test and hone their ideas. These include the American Society of Environmental History, the Association for the Study of Literature and Environment, The International Society of Environmental Ethics, The International Association for Environmental Philosophy, and the Association for the Study of Literature, Environment and Culture in Australia and New Zealand. Also, since the early 2000s, Tamkang University in Taiwan has been a hub of organising for international eco-discourse conferences. There, leading voices in the humanities, ethnography, history and geography have come together with Indigenous writers and artists, including Deborah Bird Rose, Linda Hogan, and Syaman Rapongan, among others. This has made Taiwan a key site for piloting the transdisciplinary process that we call in our pages 'integrating knowledges'. Additionally, many meetings and field trips for ethnic studies and environmental studies scholars organised by Hsinya Huang at National Sun Yat-sen University in Kaohsiung, Taiwan, have inspired crucial conversations that are reconsidered in this volume, including one of the first presentations by Michael Zimmerman of a transdisciplinary approach that he calls 'integral ecology'. This approach finds a place in Pope Francis' 2015 Encyclical, which, in turn, has been credited with having a positive influence at the 2015 Paris Climate Talks. We see the Pope's discussion of integral ecology as offering evidence that the Environmental Humanities are becoming more influential in the international sphere. We thank the hundreds of people who have organized these opportunities to meet, in North America, Australia and Taiwan. They have had an invaluable influence on the principal investigators, conveners and key researchers of H*f*E Observatories who contribute to this volume.

We warmly thank each of our contributors. Their influence in shaping the environmental humanities has already been internationally significant and we have felt incredibly honoured to work with them. We also offer our heartfelt thanks to our friends and colleagues for the stimulating conversations and insights so generously offered to us from the proposal stage to the finished manuscript stage. Elizabeth DeLoughrey, Giovanna Di Chiro, Karen N. Salt, and Kyle Whyte read proposals and introductions and provided amazingly helpful comments over the course of the entire project. Linda Hogan and Deborah Bird Rose were especially generous with their time and talents in shaping the concept of a 'widening conversation'. Their loving hearts, and passion for the Earth and all living things, beats almost palpably in this book.

As co-editors, we thank each other. It is not easy to edit a collection of essays from opposite sides of the world. One needs a sense of humour, a broad willingness to work late into the night or early in the morning, and a passion for the project that transcends the limitations of email. We also thank our families, the ones that live with us in our houses, and the ones that work with us in the Environmental Humanities. This is a field full of love, dedication and passion for the extended multispecies family inhabiting a planet increasingly threatened by human activities. We worked for all these members of our biodiverse families. A sense of urgency, relevance and care made the project a profoundly rewarding one. For that, we are deeply humbled and grateful.

Contributors

Joni Adamson is Professor of Environmental Humanities in the Department of English and Director of the Enrivonmental Humanities Initiative at the Julie Ann Wrigley Global Institute of Sustainability at Arizona State University. She is a Principle Investigator for Humanities for the Environment and lead developer for the HfE international website (http://hfe-observatories.org).

Tony Birch is the author of five fiction books, including *The Promise* (2014) and *Ghost River* (2015). He is currently the inaugural Bruce MacGuinness Fellow in the Moondani Balluk Academic Centre at Victoria University in Melbourne, Australia.

Michael Davis is a researcher with the University of Sydney and a member of the Australian Institute of Aboriginal and Torres Strait Islander Studies. He is the author of *Writing Heritage: The Depiction of Indigenous Heritage in European-Australian Writings* (Australian Scholarly Publishing, 2007).

Giovanna Di Chiro is Professor of Environmental Studies at Swarthmore College, Pennsylvania. She has published widely on the intersections of environmental science and policy, with a focus on environmental justice, community-based environmental action research and human rights.

Elizabeth DeLoughrey is Professor of English at the Institute for the Environment and Sustainability at the University of California, Los Angeles. She is co-editor of many books including *Routes and Roots: Navigating Caribbean and Pacific Island Literatures* (2007) and a forthcoming book about climate change, empire, and the literary and visual arts entitled *Allegories of the Anthropocene*.

Linda Hogan (Chicaza) is an internationally recognised poet, novelist, writer and public speaker. Her books of poetry include *DARK. SWEET: New and Selected Poems* (2014), *Rounding the Human Corners* (2008) and *The Book of Medicines* (1993). Her novels include *People of the Whale* (2008), *Solar Storms* (1995) and *Power* (1998).

Hsinya Huang is Professor of American and Comparative Literature at National Sun Yat-Sen University in Taiwan. She is the author or editor of many books and articles on Native American and Indigenous literatures. She is founder of the 'Island and Ocean Ecology Research Network' and the Convener of the Asia–Pacific H*f*E Observatory.

Mark Klett is a photographer and Regents Professor and Distinguished Sustainability Scholar at Arizona State University. He has received fellowships from the Guggenheim Foundation, the National Endowment for the Arts, the Japan/US Friendship Commission and the Pollock-Krasner Foundation.

Tyrone Martinsson is Senior Lecturer in Photography at the Valand Academy, University of Gothenburg, Sweden. His work on the Arctic includes *Rephotography: A Dialogue with History in an Arctic Landscape*, *Expedition Svalbard – Lost Views on the Shorelines of Economy* and *Arctic Views: Past and Present* (Art &Theory Publishing).

Syaman Rapongan is a prominent Tau aboriginal writer in Taiwan. His notable works include *The Face of a Navigator* (2007), *The Old Seaman* (2009), *The Eyes of the Sky* (2012) and his most recent memoir *Drifting Dreams on the Ocean* (2014). His works have been translated into English, French, Japanese, Spanish and German.

Kate Rigby FAHA is Professor of Environmental Humanities at Bath Spa University and Adjunct Professor at Monash University, Melbourne, Australia. She was the inaugural President of the Association for the Study of Literature, Environment and Culture (Australia–New Zealand) and is currently a key researcher with the Humanities for the Environment Australia–Pacific Observatory.

Deborah Bird Rose is an adjunct Professor in Environmental Humanities at the University of New South Wales, Australia and a Fellow of the Academy of Social Sciences in Australia. She is a founding co-editor of the journal *Environmental Humanities* and author of *Wild Dog Dreaming: Love and Extinction* (2011) and *Dingo Makes Us Human* (2009).

Karen N. Salt is an Assistant Professor at the University of Nottingham, where she co-directs the Centre for Research in Race and Rights and leads Europe's first-ever Black studies PhD programme. She is a key researcher with the Humanities for the Environment Asia–Pacific Observatory. Her book *The Unfinished Revolution: Haiti, Black Sovereignty and Power in the Nineteenth-Century Atlantic World* is forthcoming from Liverpool University Press.

Dr Kirsten Wehner is Head Curator, People and the Environment, at the National Museum of Australia. She has produced over twenty exhibitions, digital interactives and multi-media projects exploring Australian history, and is the co-editor/author of *Curating the Future: Museums, Communities and Climate Change* (Routledge, 2016).

Kyle Whyte holds the Timnick Chair in the Humanities at Michigan State University. He is Associate Professor of Philosophy and Community Sustainability and a faculty affiliate of American Indian Studies and

Environmental Science and Policy programmes. He is an enrolled member of the Citizen Potawatomi Nation and serves on the U.S. Department of Interior's Advisory Committee on Climate Change and Natural Resource Science.

Michael E. Zimmerman taught philosophy at Tulane University and the University of Colorado, Boulder, for 41 years before retiring in 2015. He is the author of four books and more than 100 scholarly articles and book chapters.

Part I

Integrating knowledge, extending the conversation

1 Introduction

Integrating knowledge, forging new constellations of practice in the environmental humanities

Joni Adamson

> Something deeper than just a 'shift' has to take place in our time. It may come from the backbone of human spirit and its felt connection to the earth.
>
> Linda Hogan (Chicaza), 'Backbone: Holding Up Our Future'[1]

> Country [is] a being (not a 'thing') and 'caring for country' … a practice (not a 'policy') … it consists of people, animals, plants … the dead, and the yet to be born … There is sea Country and land Country; sky Country too … it exists both in and through time.
>
> Deborah Bird Rose, 'Country and the Gift'[2]

Something deeper than a shift

At a ceremony held on Earth Day, 22 April 2016, 177 parties signed the Paris Climate Change Agreement negotiated at the 21st meeting of the Conference of Parties to the United Nations Framework Convention on Climate Change.[3] The record number of signatories seemed to signal something of a shift towards action on climate change that might result in a more socially equitable and sustainable future. Leading into the Paris talks, Pope Francis of the Catholic Church published an encyclical, *On Care for Our Common Home: Laudato Si´* widely said to have influenced the outcomes of the meetings. Anticipated by the press as a statement on climate change, the Pope, in fact, addresses climate change only briefly in his lengthy document. Instead, like the framers of 'The Future We Want,' the outcome document of the 2012 Rio+20 United Nations Conference on Sustainable Development, the Pope takes economic development and consumerism on directly, arguing that accelerating environmental change is a symptom of attitudes and practices of an unconstrained global capitalism. He charges Catholics (and members of other theistic traditions) with an obligation to care for 'Creation,' the term he uses for the Earth. Over twenty years earlier, the more secular *Our Common Future* (known as the Brundtland Report) published in 1987 by the United Nations World Commission on Environment and Development was among the first international documents to discuss the environment and global economic development together. It is widely credited with having introduced notions of 'sustainable' economic development into international consciousness. Both 'The Future We Want' and 'On Care for our Common Home' reframe 'sustainable development' as an issue involving more than economics. There is growing consensus that 'sustainability' must address not only economics, but justice and equity, and emphasise the importance of human values, attitudes, imagination and both cultural and biological diversity.

Signed by all UN member states, 'The Future We Want' confirms a broad general agreement that global society should strive for a high quality of life that is equitably shared and sustainable for all species, not just the human, especially now, when many global and community leaders are confirming that we live in an age of rapidly accelerating social disparities being exacerbated by environmental factors.[4] It affirms the efficacy of science-oriented explanations of environmental change but calls for change in the ways that research is done in international-level research programmes such as the United Nations Intergovernmental Panel on Climate Change (IPCC) and Future Earth.[5]

While it is well recognised that global efforts to monitor, measure and reverse the biogeophysical drivers of climate change will require science, many scientists and international research bodies are acknowledging that the analytical power of science stops short of investigating the main driver of planetary change – the human factor (Holm et al. 978). What humans believe and value, how they organise themselves and what they are willing to invest to achieve their goals are factors that lie largely outside scientific calculation. 'One hard lesson that we have learned from the first five IPCC assessments,' writes humanist Steven Hartman in a blog for Future Earth, 'is that rising scientific confidence and consensus do not in themselves produce shifts in societal values and norms, or changes in human behavior on a significant scale' (Hartman).

As discussed by philosopher Michael E. Zimmerman in Chapter 10 of this volume, the 2015 encyclical by Pope Francis offers evidence that religion, long a shaper of human values, long a focus of the humanities, and long predicted by some as a human occupation that would be relegated to the dustbin of history, retains an important place in global affairs. This suggests that rumours that the humanities are becoming less relevant in higher education, a popular topic in the global media, are false. Statistics show that the number of students taking degrees in the disciplines traditionally classified as the humanities have remained constant since the 1980s (Adamson 'Humanities' 136; Nye et al.; Bérubé; Paul and Graff). Moreover, as contributors to *Humanities for the Environment (HfE): Integrating Knowledge, Forging New Constellations of Practice* discuss, in the first decade of the twenty-first century, scientists, policy-makers, business and education leaders are declaring the 'environmental humanities' crucial to addressing the anthropogenic factors contributing to dramatic environmental changes.

The environmental humanities include history, philosophy, aesthetics, religious studies, literature, theatre, film and media studies informed by the most recent research in the sciences of nature and sustainability. The chapters in this collection illustrate why there is increased attention being paid to this field. Contributors explore the ways in which humanists, in collaboration with artists and community leaders, or in their own pedagogical or research practices, are responding to warming oceans, bleaching coral, disappearing islands, collapsing fisheries, melting glaciers, evaporating reservoirs of water, dying deserts, exploding bushfires, spreading radioactive contamination, and accelerating extinction rates, as well as migrating populations of people, plants and animals.

Humanists have long been researching and writing about changing environmental processes that are 'inescapably entangled with human ways of being in the world' (Rose et al. 1). As I and many others concerned with environmental justice and postcolonial ecologies have explained, the entangled disciplinary fields that find confluence in the environmental humanities are arguably much older than the early 1990s dates usually given to the emergence of environmental literary criticism, history and philosophy (Adamson and Ruffin 5–12; DeLoughrey and Handley 9–20). Building on these long histories, 'environmental humanities,' as a term, was first suggested in Australia in 2001 by historian Libby Robin, ethnographer and literary critic Deborah Bird Rose and ecofeminist Val Plumwood, who formed a group to study the 'ecological humanities' (Nye et al.). Others who have worked for two decades or more in ethnic studies, Indigenous studies, environmental justice and postcolonial studies, political ecology, and religion and ecology, have also argued that the roots of the environmental humanities may be traced back to some of the earliest cosmological narratives, stories and symbols among the world's oldest cultures. Many cosmologies tell not only of cultural origins, they sometimes reference uprisings, resistance or revolts in the colonial world, or suggest critiques of imperialism and the enforcement of Western religions on colonised peoples (Adamson 'Humanities' 136). Examples include Greek oral poet Hesiod's accounts of Pandora's jar *and* unjust Greek appropriation of resources and capture of slaves (ca 700 BCE), North American Indigenous creation stories *and* a revolt against Spanish conquistadors in 1620 CE, and the first printed versions of the Mayan creation story, the Popol Vuh, with its account of the actions of multiple gods *and* its suggestion of opposition to the Catholic Church's notions of *one* god (Hesiod ll. 320–341; Adamson *American Indian* 29; Tedlock 263–264, 270–271). Each of these cosmological stories illustrate why many environmental humanists are arguing that climate change is only the latest development of a history of colonialism inflicting anthropogenic (human-caused) harms on the world's Indigenous and ethnic minority peoples. To borrow words from contributor and philosopher Kyle Whyte, stories about climate change suggest something like déjà vu for the world's Indigenous peoples, since recent anthropogenic environmental change continues patterns 'that have been part of settler colonialism for quite some time' (Whyte, Chapter 7).

While stereotypes associated with humanities scholarship (dry discourse analysis, arcane debates) may have once made these disciplines seem ill suited to addressing crises outside the walls of academe, Whyte's analysis illustrates that many twenty-first century humanists are creatively re-directing analysis of ancient and modern texts, in order to contribute to a rethinking of the 'human factor' involved in climate injustices and other complex social and ecological challenges. In a succinct essay examining the intellectual history of work to deconstruct culturally-constructed binaries between 'Culture' and 'Nature,' Dianne Rocheleau and Padini Nirmal trace the split that began appearing between the humanities and the sciences in the mid-twentieth century when influential British physical chemist and novelist Charles Percy Snow, in a lecture at Oxford in 1959, argued for what he termed a 'two culture' divide (50–54). Snow buttressed increasingly unquestioned assumptions about the reasons modern cultures should shift their focus towards the highly specialised and disciplined methods of knowledge production categorised as 'science,' which had long been used to authorise the terms of colonial expansionism and provide a rationale for neocolonial development schemes (such as dam construction, deforestation, mining and oil extraction) (Rocheleau and Nirmal 50–54; Adamson 'Humanities' 136–138; Latour 479–480; Snow).

From the 1970s forward, humanists began forming professional organisations to heal intellectual divides and 'turn' their disciplines towards more sophisticated engagement with environmental issues. These included the American Society of Environmental History (1977), The International Society for Environmental Ethics (1990), the Association for the Study of Literature and Environment (1993), and its 'sister organisations' (Europe, Japan, India, Australia, New Zealand and Taiwan), the International Association for Environmental Philosophy (1997), the Forum on Religion and Ecology (1998) and the Australian Environmental Humanities Hub (2001). Several humanities scholars have commented on the 'coterminous turn' of these organisations, at the beginning of the twenty-first century, not only towards energised activity to 'heal the two culture split' but towards discussion of the 'Anthropocene' (DeLoughrey 352; Di Chiro). This neologism was proposed in a short essay published in 2000 by Paul Crutzen, a Nobel-prize winning atmosphic chemist, and Eugene Stoermer, a biologist, who raised questions about whether or not the Earth had entered a 'new Age of Man,' in which *anthropos*, the Greek word for human, had become like 'Nature' itself, a geomorphic force so powerful that – as a species – they were changing every biogeochemical system on the planet (Crutzen and Stoermer). Members of the Anthropocene Working Group of the Geological Society Stratigraphy Commission, formed to study the scientific validity of the term, have noted that although 'Anthropocene' has still not gained full approval among geologists or the scientific community at large, it is widely becoming a useful meta-concept for describing the biogeophysical forces catalysed by human activities. If accepted by scientists, it would be used to mean that the Earth has transitioned from the Holocene, the epoch that began at the end of the last Ice Age, approximately 11,500 years ago, to a new geological epoch, the Anthropocene or 'Age of the Human' (Zalasiewicz et al. 14–16).

Since the publication of Crutzen and Stoermer's essay, scholars who focus their research on the environment, and in advance of full scientific approval of the term, have been pouring 'in their interdisciplinary masses' into symposiums and conferences to discuss and debate this 'ephochal idea' (Nixon). Over the last fifteen years, there has been a dramatic shift away from unquestioned assumptions about science, and an increasing number of calls for the inclusion of humanities perspectives in environmental decision-making. This has spurred expansion of the institutionalisation, infrastructural support, funding and networking in the environmental humanities.[6]

One of the most ambitious projects to emerge from this activity, 'Humanities for the Environment,' or H*f*E, undergirds the work in this volume. Funded by a generous grant from the Andrew W. Mellon Foundation to the Consortium of Humanities Centers and Institutes (CHCI), the initial phase of the project was organised by three universities – The University of Sydney (Australia), Trinity College Dublin (Europe) and Arizona State University (North America) – that, in turn, worked to expand and network other universities and programmes developing the environmental humanities internationally.[7] At one of the first H*f*E events, a conference in 2014 called Encountering the Anthropocene: The Role of the Environmental Humanities and Social Sciences Conference, held in Sydney, Australia, Sverker Sörlin, an historian of ideas and environment at Royal Institute of Technology in Sweden, observed in a keynote speech that humanists and social scientists are very aware that 'what we do now' will have 'consequences for a long time' (Sörlin n. pag.).[8]

This sense of urgency, relevance and need for action among a fast growing number of humanists and social scientists helps to explain why the H*f*E project was purposefully designed to 'network the networks' of professional organisations, initiatives and programmes. With the Andrew W. Mellon Foundation as major funder, H*f*E conveners and researchers were guided by the Foundation's persistent defence of the value of the humanities and its consistent encouragement of 'scholars and institutions to experiment and adapt' (Howard 4). To this end, the universities piloting the first phase of the project organised themselves into a system of 'Observatories'. The 'Observatory' was chosen as the formal mechanism that partner institutions would use to network with other universities and environmental humanities initiatives around the world. A common website, considered one of the major deliverables of the Mellon-funded grant, was designed to employ digital tools to network humanities centres and institutes engaging in environmentally-focused projects and explore how digital tools might be used to tap the creative potential of new collaborations emerging among humanists, social scientists, artists, scientists, business leaders, government officials and engineers (Holm et al.).[9]

Each university working in the H*f*E network was charged with organising activities and programming, loosely integrated with those of their partner institutions and universities, and, at the same time, focusing on regionally distinct issues, themes, audiences, methods and arguments. A set of 'Common Threads,' found on the international H*f*E website, guides these distinct but connected projects.[10] The first Common Thread states: 'We hold that … the humanities – are a largely untapped resource of insight into human motivation and agency that should inform how we interpret and stimulate change.' Each H*f*E Observatory would also have a common goal to alert widening publics to the extraordinary 'role of humans' in 'transforming the chemical, physical and biological processes of the Earth's atmosphere, land surfaces and oceans at an ever-increasing pace'.

In what follows, I briefly introduce the organisation and major themes of *Humanities for the Environment (HfE): Integrating Knowledge, Forging New Constellations of Practice* and elaborate on the ways that the HfE 'observatory' system is networking researchers to build integrated knowledges and model new 'constellations of practice' in the humanities. I also introduce a dual-chapter 'conversation' found in the first section of the volume between famed Chicaza (North American) poet and novelist Linda Hogan and founding environmental humanist Deborah Bird Rose. After these opening conversational chapters, in Chapter 4, co-editor Michael Davis introduces Parts II and III and elaborates on the ways each chapter is connected to the concepts of 'Backbone' (Hogan, Chapter 2) and 'Country' (Rose, Chapter 3) and brings the volume into conversation with the concept of the Anthropocene. The aim of this collection as a whole, to use the lyrical words of Hogan and Rose, is to seek something 'deeper than a shift' in human perception (Chapter 2), something that might be described as 'encounter, not division' and respectful attention not only to diverse scientific literacies and knowledges, but to what we all have learned from where we have come and where we are going (Chapter 3).

The HfE observatory system

At the highest levels of global research activity on changing environments, including the influential IPCC and Future Earth, there is a growing clamour for 'change in the *intellectual* climate' (Castree et al.). The HfE project taps into this sense of urgency through the use of the term 'observatory'. This descriptive word was chosen to quicken the imagination of humanists being called upon to think outside the limitations of traditional humanities research protocols, such as the single-authored monograph or narrowly disciplinary essay, and to engage in more collaborative, transdisciplinary or digital projects and research across all the disciplines required to understand both social and natural systems. The word 'observatory' also aligns well with research platforms and scientific initiatives such as Future Earth by evoking a sense of a humanities 'laboratory,' or 'research space' that would offer humanities points of view. Each of the contributors to *Humanities for the Environment (HfE): Integrating Knowledge, Forging New Constellations of Practice* is networked to the others through the North America Observatory (Arizona State University), the Asia–Pacific Observatory (National Sun Yat-sen University and National Chung-hsing University) and the Australia–Pacific Observatory (University of Sydney). These observatories are purposely organised to reflect and contribute to the interdisciplinary research being conducted at Future Earth and other international research platforms.

The word 'observatory' is also inspired by diverse human groups, in many places around the world, who, over the course of hundreds, even thousands of years, have been renowned for their remarkable achievements in art, astronomy, landscape design, architecture, storytelling, mapping, social and economic organisation, and engineering. For example, the ancient Hohokam peoples of North America, discussed by Joni Adamson in Chapter 8, lived in harsh desert climates and, from roughly 400 BCE to 1450 CE, built remarkable complexes of observatories as instruments for collecting information to help them better understand the stars and seasons, and to build water control features like dams, canals and headgates for improving their agricultural outputs and social well-being. Today, modern Phoenix, Arizona, sprawls around one of the largest ancient ruins of a Hohokam city, called 'Pueblo Grande' or 'Big House'. Pueblo Grande is only one example of the many pyramidal-sub-structured mounds that can be found throughout Central and North America, from the Maya pyramids in Oaxaca, Mexico, to the Mississippian culture mounds at Cohokia in Missouri discussed by Linda Hogan in Chapter 2. The descendants of the Hohokam, known today as the O'odham, meaning 'People,' continue to live on their own lands in and around Phoenix, a megacity of four million people, where they are reintroducing traditional foodways and patterns of agriculture that are suggesting how modern city dwellers might look to an 'ancient future' as they build equitably and sustainably for tomorrow (see Adamson 'Ancient Future').

HfE Observatories are also inspired by other ancient peoples, such as the Austronesian-dialect speaking groups that Hsinya Huang and Syaman Rapangan introduce in Chapter 11, who inhabit much of the Asia–Pacific region. They are thought to have begun migrating from island to island approximately 6,000 years ago using remarkable navigational skills and geographical knowledge they derived from using a 'star compass,' a kind of mnemonic device that was taught through the use of circular or rectangular diagrams where each of 32 compass points indicates the rising or setting position of guide stars, the flight path of birds and the direction of the ocean's waves.[11] Unlike European charts, the position of islands on these mental compasses is relative to their central point: the island or canoe on which the navigator is sitting. The navigator only knows where he or she is in this kind of navigation by memorising the point from which he or she sailed. This calls upon the paddler to engage in 'constant observation' (Thompson).

In Taiwan today, some Austronesian groups continue to criss-cross the Pacific using the same cosmological, cartographic, observational and mnemonic skills that have been thousands of years in the making. One of these groups, the Tau, live on a small volcanic island named *Pongso-no-Tau* (literally, Island of Original People), also known as Orchid Island, located 40 kilometres southeast of Taiwan (Tsai). As contributor Syaman Rapongan, a Tau elder, has written in his memoir *Drifting Dreams on the Ocean*, the Tau continue to practise their long-honed fishing and foodways and techniques of canoeing and boat-building. At the same time, Rapongan is using his observational and navigational skills to seek partnerships in a widening anti-nuclear, anti-militarism movement that is not focused solely on Hiroshima and Nagasaki but on nuclear contaminations across the Pacific, including that of his own island, *Pongso-no-Tau* (Chapter 11).[12] Thus, the word 'observatory' evokes not only recent international research platforms such as Future Earth, but over 10,000 years, to use Michael Davis' words from Chapter 4, of human observing, seeing, reflecting and knowing at diverse ancient and modern sites, on land and sea. At the same time, a fundamental tenet embraced by the HfE Observatory system is that 'observation' is never neutral, and that values, histories, archives, theories and affects, and ideologies and instruments provide both the ground and the frames for what we 'see' or do not see – and make available for others.

For over a decade before the launch of the HfE project in 2013, these diverse notions, cultures and histories were inspiring researchers who would later organise through the HfE project to study the links and convergences between Native North American, Indigenous Pacific, Australian Aboriginal and Torres Strait Islander, Caribbean and Asian–Pacific Austronesian Studies. Many of the researchers building these networks have already been piloting decades-old frameworks of environmental justice, postcolonial and new material ecological studies that would later form the foundation for the environmental humanities (Adamson and Ruffin; DeLoughrey and Handley; Iovino and Opperman; Chang and Slovic; Huang and Chang). As Elizabeth DeLoughrey explains in Chapter 16, the humanities has tended to focus on diverse interdisciplinary engagements with the representations of territorially-based, geophysical place. She argues that the oceanic turn has arisen from Cold War geopolitics, a shift in the interdisciplinary study of space as a practiced place, and the rise of diaspora and globalisation studies. Also, twenty years ago, after Tongan anthropologist Epeli Hau`ofa published his deeply influential essay titled 'Our Sea of Islands' arguing that the legacies of colonial belittlement that render the Pacific as 'islands in a far sea' needed to be replaced with a more accurate and world-enlarging view of the complexity of islands and the primacy of the oceans, humanists began paying more attention to the entangled connections between land and sea. Hau`ofa could not have foreseen, DeLoughrey observes, the ways in which climate change, particularly sea-level rise, would further transform island and ocean studies and shift focus in the environmental humanities from a long-term concern with mobility and fluidity across biotic, regional, national and colonial boundaries towards theorising ways of embedding, animating and submerging, rendering islands and oceans into recognised places (see Chapter 16; and Davis, Chapter 4).

After the launch of the HfE project, these shifting concerns and expanding networks would greatly influence the themes articulated by each observatory for individual and linked research, meetings and projects. The North America Observatory, under the thematic framework 'Building Resilience in the Anthropocene,' set about rethinking relationships between human beings and diverse knowledge systems and between humans and other species. Under the thematic framework, 'Between Land and Sea,' the Asia–Pacific Observatory began networking scholars from throughout Southeast Asia, North America and Europe studying the trans-oceanic cultures and ecologies of Austronesia and Taiwan, the Caribbean and the Pacific. The Australia–Pacific Observatory brought the concepts of the 'Anthropocene' and 'Country' into conversation. 'Country,' in the words of Deborah Bird Rose, articulates a unique tradition of Australian Indigenous ecological philosophical and practice developed over many hundreds of thousands of years; it is 'sea Country and land Country … sky Country too'. As explained in the next section, the concept of Country, discussed by Rose, and another concept, Backbone, discussed by Hogan, became the organising terms for this volume which presents the work of H*f*E researchers from all three observatories.

Extending the conversation, integrating knowledges

In launching *Environmental Humanities* in 2012, Deborah Bird Rose observed that the aim of the new scholarly journal would be to provide a space for a 'wide range of conversations' that needed to take place in this 'time of growing awareness of the … challenges facing all life on Earth' (Rose et al. 1–5). This first part of *Humanities for the Environment* (H*f*E)*: Integrating Knowledge, Forging New Constellations of Practice* aims to contribute to a widening conversation by bringing Linda Hogan and Deborah Bird Rose into dialogue. In the first chapter, 'Backbone: Holding Up Our Future,' Hogan argues that in times of great danger, something deeper and more lasting than a 'shift' in human perception is needed. What is required is a felt connection to the earth, one that taps into the 'Backbone' of the human spirit. To explain what she means by 'shift,' Hogan references her novel *Solar Storms* (1995) which fictionalises events surrounding the construction of a hydroelectric complex of dams in James Bay, in northern Quebec, Canada. Based on the resistance of the Cree and Inuit peoples to this development, the novel depicts characters with a clear understanding, *before* the reservoirs are filled, of the cultural and biospheric changes that will take place. Indeed, the Cree were devastated by the loss of their caribou herds, as about twelve thousand animals drowned when the reservoir behind the first dam began filling.

Today, twenty years after the publication of *Solar Storms*, in *People of a Feather* (2011), filmmaker and biologist Joel Heath documents how hydroelectric energy development in Canada sets in motion processes that have literally changed the biogeochemical currents of water and air at a planetary scale. In the Belcher Islands of Hudson Bay, the eider duck, upon which the Inuit people depend for food and clothing is on the brink of extinction. Heath, who spent seven winters filming the consequences of dam construction on the Belcher Island Inuit, uses time-lapse photography and satellite imagery to reveal how a massive dam complex that powers New York City and other eastern North American cities has changed 'the chronology of fresh water inputs, influencing ocean currents, and sea ice habitats' (Heath n. pag.). As a result, the seasons are shifting and the ice patterns changing the eider ducks' access to food. This is leading to the extinction of the duck and the displacement of the Inuit.

Although Hogan does not write about *People of a Feather*, this brief introduction to Heath's documentary helps to provide visual and statistical evidence of the prescience of *Solar Storms* and the concept of 'Backbone' that she introduces in Chapter 2. The characters in *Solar Storms* vigorously resist dam construction because they understand that it will change not just one place, northern Canada, but the entire planet. They live in villages with names such as 'The Creator's Elbow,' 'The Creator's Knee' or 'The Creator's Eyebrow' that begin to disappear under the water when the dam begins filling. The village names once established cognitive connections between humans and what Hogan calls the 'Backbone of Creation'. The concept of Backbone provides a different visual, spiritual and cognitive order, a different kind of map, so to speak, suggesting that the Earth, 'like a human spine, has circuitry, electricity, and impulses that are active and dynamic' (Chapter 2). Backbone calls upon humans to care as much about the planet's circuitry and circulation as they do about any one specific place.

Rose enters the conversation about 'Backbone' by fleshing out a lecture she delivered at the HfE *Encountering the Anthropocene* conference. In 'Country and the Gift' (Chapter 3), she explains that 'Country,' like Backbone, signals the fleshy, embodied, living and structured quality of Earth life. Both Backbone and Country convey Aboriginal/Indigenous scientific literacies and knowledges that can still be found strongly in evidence among Indigenous peoples worldwide but which are known by different words and terms in different languages. Hogan and Rose bring these terms into conversation with the concept of the Anthropocene, as they call for richer, more culturally integrated understandings of the complex anthropogenic activities involved in climate change and accelerating extinction rates. Both chapters also flesh out another emerging area of the environmental humanities that S. Eben Kirksey and Stefan Helmreich have called 'multi-species ethnography,' an important methodology that is calling upon humans to radically rethink life and the ways in which all sentient beings, including humans, live in 'situated connectivities that bind us into multi-species communities' (Kirksey and Helmreich 546).

At the same time, both Hogan and Rose urge readers to engage with differing ontologies, seriously and lovingly. We must be respectful, Rose observes, and willing to open ourselves fully to ideas much older than recent scientific terms such as the 'Anthropocene' that focus largely on *anthropos*, or the human (Rose, Chapter 3; see also Davis, Chapter 4).

Other contributors, including Kyle Whyte and Anthony Birch, also enter this conversation about Anthropocene discourses. They caution against terminology that relegates Indigenous, Aboriginal or ethnic minority groups to a state of *non-*existence, pulling them into a universal 'We' that results in the intellectual equivalent of the *terra nullius* narrative of the late eighteenth and nineteenth centuries (Birch, Chapter 13). This mythology, embraced by Euro–Austro–American settler colonials on many continents, continues to allow neocolonial powers to mask their histories of violence. Also, concepts emerging from diverse ontologies, such as Backbone and Country, can help expand understanding of the aims and goals of international documents like The World Peoples Conference on the Rights of Mother Earth and Climate Change held in 2010 in Bolivia. This document perceives the Earth as a sentient entity with circuitry, electricity and impulses that are active and dynamic. The Declaration directly confronts a universal *anthropos*, or 'We' found in early documents outlining sustainability such as the Brundtland Report that tend to confer a kind of homogeneity on all human experiences and fail to acknowledge differences in histories or differences in contribution (or lack of contribution) to current CO_2 levels. Delegates to the World People's Conference differentiate between a universal 'We' and the word found in their document, 'us'. They describe ongoing environmental and social destruction of the places that Indigenous and Aboriginal First Peoples live as 'assaults and violations against our soils, air, forests, rivers, lakes, biodiversity, and the cosmos' as 'assaults against us' (Indigenous Peoples Declaration, Preamble n. pag.). 'Us' is emphasised to mean humanity in all its differences, and to include all living beings, not just humans. Instead, 'us' is interpreted to mean whole ecologies, whole forests, whole riverine systems, with all their situated connectivities. Other legislative instruments, such as Ecuador's Constitution, revised in 2008, to include Title VII, often referred to as the 'Bien Vivir' or 'good way of living chapter,' acknowledge Indigenous ways of knowing that are expanding the concept of human rights to include rights for the intergenerational, evolutionary space and time required not just for the survival of humans but for all species (Constitution of Ecuador, 2008, Title VII, Chapter 2, Secs. 1–7, 134–139; also see Adamson 'Whale as Cosmos').

Backbone, Country and notions of situated multi-species connectivities also help to explain why HfE researchers, as affirmed in the Common Threads, 'study intertwined human and ecological histories' and 'recognise various ways of knowing, including place-based and Indigenous knowledges, as crucial to developing new approaches to human purpose and motivation'. This advocacy of integrated understandings of deep time, evolution and the reasons why we should be concerned with 'the not yet born' (Rose, Chapter 3) also contributes to emerging concepts of 'intergenerational justice' not just for *anthropos*, but for all species now found in documents that range from 'The Future We Want' to the recent Paris Agreement on Climate Change (Rose 'Introduction'; Di Chiro; DeLoughrey; Haraway).

Contributors to *Humanities for the Environment (HfE): Integrating Knowledge, Forging New Constellations of Practice* ask readers to think about how 'we,' in all our intergenerational, multi-species diversity, might connect the 'backbone of human spirit to its felt connection to the earth' and drive shifts in societal values and norms that catalyse interventions in local, regional, and global governance (Hogan). Many academics are arguing that to do this, we will need the full force of transdisciplinary alliances between humanists, social scientists, natural scientists, artists and faith-based and secular community members who are seeking not one plan of action, but a range of evidence-based, reasoned, scaled and culturally diverse responses 'reflective of life in a plural world' (Castree et al. 765–766). This suggests why influential thought leaders on the stage of climate change, including Pope Francis, have been welcomed as they help to scale up the numbers of people who might join new alliances of 'us' (see Zimmerman, Chapter 10).

Pope Francis's encyclical raises questions such as 'How might we acknowledge the sheer scale of the challenges we face while also appreciating the intimate resonance of individual and multi-species communal experience?' To answer questions such as this, each of the international documents cited above envision a shift in scale that anticipates the 10 billion people projected to be living on the planet in a near future. To meet this challenge, Hogan argues, we will need to 'quit haggling over singular words or concepts [the Anthropocene]' and 'get to work ending the threats to our environments' (Chapter 2). This call to 'get to work' resonates in each chapter in the volume, as contributors examine the ways that humanists are innovating 'new constellations of practice' in their disciplines and in their communities.

New constellations of environmental and digital humanities practice

A high percentage of Andrew W. Mellon Foundation dollars has been granted to support the environmental and digital humanities and to make project outcomes more 'openly available' (Howard 9, 10–11). For this reason, one of the major deliverables of the Mellon Foundation grant to the H*f*E project has been the creation of a common international website designed to employ digital tools to 'network the networks' of humanities centres and institutes and to craft new 'constellations of humanities practice'. None of the tools or practices employed by H*f*E researchers is, in itself, new. What is 'new' is the ways in which these methodologies and practices are being 'constellated' by collaborative teams piloting research and projects employing long-tested humanities practices. 'Constellations of practice' is an apt phrase for H*f*E Observatories generating methodologies and practices that are seeking to move beyond traditional contemplative or reflective humanities outcomes, and work for solutions to some of the greatest challenges of our time. As the lead developer of the H*f*E international website team, I have chosen the word 'constellation,' not only because of my own longtime scholarly interest in ancient observatories, almanacs and oral narratives as 'seeing instruments' but also because it is associated with astronomy, constellations and the stars to which most human cultures have linked their founding cosmologies (Adamson 'Humanities' 136–138).[13]

With this in mind, H*f*E researchers have been piloting projects that aim to adapt and experiment with the humanities while keeping and valuing best practices. They are asking questions that humanists have always cared about, including '*How* do the humanities and the arts contribute uniquely through storytelling to solving social justice and environmental challenges?' 'Can the humanities catalyse imagination of new ideas, narratives, frameworks, alternatives, demands, and projects that will enable people to envision plausibly different, even "livable futures"' (Adamson 'Humanities' 139)? In Chapter 6, Giovanna Di Chiro restates these questions by asking how those teaching in the environmental sciences and humanities might create more active, embodied, collaborative and *life-affirming* curricula, theories and community engagements by asking students to move beyond declensionist narratives and re-imagine the meanings of survivability, thriveability and the longue durée. Through the power of story, creative nonfiction and memoir, Di Chiro 'scales up' the outcomes of her teaching and community work and even experiences joy as she and her students think about how we might live convivially together or, at the very least, forestall or change the litanies of trauma, disaster and extirpation that we face every day in news of a fast changing planet.

Di Chiro's discussion of the creation of green jobs also conveys how 'action research-oriented' pedagogy can help face down 'hopelessness' and inspire a kind of 'soular-solar' or life affirming activeness, that in the words of Linda Hogan, encourages us to 'get to work' on the climate injustice and poverty challenges we are facing. Di Chiro's socially embedded pedagogy also connects her discussion, and the environmental humanities, to Pope Francis' suggestion that 'green jobs' might lead to new forms of 'social uplift sustainability'. Other HfE researchers are also 'scaling up' through the creation of pedagogical templates that are made easily available on digital platforms, such as the HfE international website. As I explain in Chapter 8, teachers and students are developing multimodal courses that can be replicated, adapted and scaled for multiple locations and diverse groups of students. Reflecting the goals of the HfE Observatories, I examine how a campus can be turned into an urban laboratory where students pilot new forms of 'citizen humanities' that complement and expand the work of citizen scientists creating plausible 'futures we want'.

Contributors also employ history or narrative as the 'thread' that ties together new constellations of photography, rephotography, curation, interpretation, mapping, modelling and visualisation. In Chapter 9 photographers Mark Klett and Tyrone Martinsson address time, change and perception of climate change in photographs repeated at different times. Through the visual technique of rephotography Klett and Martinsson document physical locations in the Western United States and in Svalbard in the Arctic, where they record evaporating reservoirs of water or receding glaciers. In Chapter 15, Kirsten Wehner discusses how museum curators are re-shaping their practices in the context of the Anthropocene, and seeking to enable people already affected by changes to envision their own plausible futures.[14]

Seeing, observing, moving, walking and paddling together

Contributors to *Humanities for the Environment* (HfE) explore new constellations of humanities practice that seek to transform social values, expand understanding of foundational principles of justice and sustainability, rewrite inaccurate narratives about human relationships to ecosystems and nonhuman species, and recalibrate strategies for adaptation to changing needs and conditions. As Michael Davis explains in Chapter 4, each chapter in the collection expands the conversation about Backbone and Country into an exploration of new forms of humanities 'seeing, observing, moving, walking and paddling toward knowledge'. This notion informs Davis' own chapter (12) about Aboriginal encounters with British explorers in the nineteenth century. Davis suggests these interactions can be seen as gesturing towards a collaboration that offers insight into how contributors to the volume are expanding understanding of the place of the humanities in scaled up responses to environmental change. As contributors illustrate, humanists are seeking to shift common human perceptions about human relation to the environment by tapping into what Hogan refers to as the Backbone of human spirit.

Finally, like 'Humanities for the Environment – A Manifesto for Research and Action,' a recently published proposal guiding what H*f*E researchers are calling the second, or '2.0 phase' of the H*f*E project, each chapter of this book can be read as an invitation to join H*f*E researchers in an 'open global consortium of humanities networks and organizations' (Holm et al.). Now expanding beyond its original funding and its first three observatories to new locations in Asia, Africa, South America, and the Arctic, the H*f*E network is continuing to explore the crucial ways humanistic disciplines may be constellated, on Backbone, and in Country, to 'hold up the future' (Hogan, Chapter 2).

Notes

1 See Chapter 2, this volume.

2 Lecture, 'Country and Gift', *Humanities for the Environment*. Media. February 2014. http://hfe-observatories.org/media/deborah-bird-rose-country-and-the-gift/

3 See 'Record: 177 Parties Sign', www.cop21.gouv.fr/en/a-record-over-160-countries-expected-to-sign-the-paris-agreement-in-new-york-on-22-april-2016/ As of 3 August 2016, there are 180 signatories to the Paris Agreement. See http://unfccc.int/paris_agreement/items/9444.php

4 See 'The Future We Want', https://sustainabledevelopment.un.org/futurewe want.html

5 Future Earth is sponsored by the United Nations Environmental Programme, the International Council for Science and the International Social Science Council.

6 Universities developing environmental humanities initiatives, degrees, programmes or certificate include Arizona State, Linköping University (Sweden), Princeton, National Sun Yat-sen (Taiwan), New South Wales (AU), Oxford (UK), Stony Brook, Stanford, Sydney (AU), Tamkang University (Taiwan), Trinity College (Dublin), UC, Los Angeles, UC, Davis, and the University of Utah. Networked collaborations between these universities, and many others around the world, are also catalysing the organisation of more informal networks such as the Transatlantic Research Network, Environmental Humanities for a Concerned Europe (ENHANCE) that brings the Rachel Carson Center (Munich) together with Leeds University (UK) and the KTH Royal Institute of Technology's Environmental Humanities Laboratory (Sweden) to offer doctoral degrees. For links to most of these programmes and initiatives, see Steven Hartman, 'Unpacking the Black Box'. Also see Nye et al.

7 The North American Observatory, headquartered at Arizona State University, also networked closely with Clark University and Wake Forest University. For more on the CHCI and this Mellon grant, titled Integrating the Humanities Across National Boundaries, see the CHCI website, http://chcinetwork.org

8 Four of the contributors to this volume, Deborah Bird Rose, Giovanna Di Chiro, Kate Rigby and Kirsten Wehner, were participants at this conference.

9 See the Humanities for the Environment international website, http://hfe-observatories.org

10 The HfE Common Threads were written one year into the grant period by HfE Conveners, Principle Investigators and Key Researchers who met at the 2014 international meeting of CHCI in Hong Kong. See http://hfe-observatories.org/common-threads/

11 See 'The Art of Navigation', Cultural Survival website, www.culturalsurvival.org/publications/cultural-survival-quarterly/none/art-navigation

12 See also Tsai, 'Sustainability of Small Islands in Taiwan'.

13 The H*f*E web designer and graphic artist is Patricia (Patty) Ferrante.
14 Klett and Martinsson's photographs, shown in black and white in Chapter 9 of this volume, can be seen in full color on the HfE website: http://hfe-observatories.org/projects/environmental-rephotography/. Also for more on museum curation as a new constellation of practice in the humanities, see the forthcoming volume in Routledge's Earthscan series, entitled *Curating the Future: Museums, Communities, and Climate Change*, edited by Libby Robin and Kirsten Wehner.

Works cited

Adamson, Joni. *American Indian Literature, Environmental Justice, and Ecocriticism: The Middle Place*. Tucson: University of Arizona Press, 2001. Print.

——'The Ancient Future: Diasporic Residency and Food-based Knowledges in the Work of American Indigenous and Pacific Austronesian Writers', Special Issue: 'Migrants and their Memories', Guest Co-Eds K.T. Tee, Ayeling Wang and I-Chun Wang. *Canadian Review of Comparative Literature*. 42.1 (March 2015): 5–17. Print.

——'Humanities', *Keywords for Environmental Studies*. Eds Joni Adamson, William A. Gleason, and David N. Pellow, New York University Press, 2015. 135–139. Print.

——'Whale Cosmos: Multi-species Ethnography and Contemporary Indigenous Cosmopolitics', Special Issue: 'Ecocriticism in English Studies', *Revista Canaria de Estudios Ingleses* 64 (April 2012): 13–46. Print.

Adamson, Joni and Kimberly N. Ruffin 'Introduction', *American Studies, Ecocriticism and Citizenship: Thinking and Acting in the Local and Global Commons*. Eds Joni Adamson and Kimberly N. Ruffin. New York, US and Oxford, UK: Routledge, 2013. 1–17. Print.

Bérubé, Michael. 'The Humanities, Declining? Not According to the Numbers', *The Chronicle of Higher Education*. 1 July 2013. Web.

Castree, Noel, William M. Adams, John Barry, Daniel Brockington, Bram Büscher, Esteve Corbera, David Demeritt, Rosaleen Duffy, Ulrike Felt, Katja Neves, Peter Newell, Luigi Pellizzoni, Kate Rigby, Paul Robbins, Libby Robin, Deborah Bird Rose, Andrew Ross, David Schlosberg, Sverker Sörlin, Paige West, Mark Whitehead & Brian Wynne et al., 'Changing the Intellectual Climate', *Nature Climate Change* 4 (September 2014): 763–768. Print.

Chai-ju Chang and Scott Slovic, *Ecocriticism in Taiwan: Identity, Environment, and the Arts*. London: Routledge, 2016. Print.

Constitution of Ecuador 2008, Title VII, Chapter 2, Secs. 1–7, 134–139.

Crutzen, P. and Stoermer, F. 'Have We Entered the Anthropocene?' *International Geosphere-Biosphere Program Newsletter* 41 (2000): 17–18. Print.

DeLoughrey, Elizabeth. 'Ordinary Futures: Interspecies Worldings in the Anthropocene', *Global Ecologies and the Environmental Humanities*. Eds Elizabeth DeLoughrey, Jill Didur and Anthony Carrigan. London: Routledge, 2015. 352–372. Print.

DeLoughrey, Elizabeth and George Handley, eds *Postcolonial Ecologies: Literatures of the Environment*. Oxford and London: Oxford University Press, 2011. Print.

Di Chiro, Giovanna. 'Environmental Justice and the Anthropocene Meme', *Oxford Handbook on Environmental Political Theory*. Eds Teena Gabrielson, Cheryl Hall, John M. Meyer and David Schlosberg. New York: Oxford University Press, 2015. 362–382. Print.

'The Future We Want – Outcome Document', Sustainable Development Knowledge Platform. United Nations Department of Development and Social Affairs. 2015. Web.

Haraway, Donna. 'Anthropocene, Capitalocene, Plantationocene, Chthulucene: Making Kin', *Environmental Humanities* 6 (2015): 159–165. Print.

Hartman, Steven. 'Unpacking the Black Box: The Need for Integrated Environmental Humanities (IEH)', Future Earth Articles. Web.

Hau'ofa, Epeli. 'Our Sea of Islands'. *The Contemporary Pacific*, 6.1 (Spring 1994): 147–161. Print.

Heath, Joel. *People of a Feather*. DVD. Montreal, Quebec: Sanikiluaq Running Pictures Ltd., 2011. Video.

Hesiod. 'Works and Days', *Sacred Texts*. Trans. Hugh G. Evelyn-White, 1914. Web.

Hogan, Linda. *Solar Storms*. New York: Simon and Schuster, 1995. Print.

Holm, Paul, Joni Adamson, Hsinya Huang, et al. 'Humanities for the Environment – A Manifesto for Research and Action', *Humanities*. 4.4 (2015): 977–992. Print. Web.

Howard, Jennifer, 'At Mellon, Signs of Change', *Chronical of Higher Education* (June 2014). Web.

Huang, Hsinya and Clara Shu-Chun Chang, eds, *Aspects of Transnational and Indigenous Cultures Series*. Cambridge: Cambridge Scholars Press, 2014. Print.

Iovino, Serenella and Serpil Opperman, eds *Material Ecocriticism*. Bloomington, IN: U of Indiana Press, 2014. 253–268. Print.

Kirksey, S. Eben and Stefen Helmreich. 'The Emergence of Multispecies Ethnography.' *Cultural Anthropology*. Vol. 25, Issue 4, 545–576. Print.

Latour, Bruno. 'An Attempt at a "Compositionist Manifesto"", *New Literary History* 41 (2010): 471–490. Print.

Nixon, Rob. 'The Anthropocene: The Promise and Pitfalls of an Epochal Idea', Edge Effects, a Project of The Center for Culture, History and Environment, University of Wisconsin, Madison, 2014. Web.

Nye, David E., Linda Rugg, James Fleming and Robert Emmett. 'Background Paper The Emergence of the Environmental Humanities', MISTRA: The Swedish Foundation for Strategic Environmental Research. Stockholm, Sweden, May 2013. Print.

Paul, Jay and Gerald Graff. 'Fear of Being Useful', *Inside Higher Education* 5 (January 2012). Web.

Pope Francis. *On Care for Our Common Home: Laudato Si*. Washington, DC: United States Conference of Catholic Bishops, 2015. Print.

Rocheleau, Dianne and Padini Nirmal. 'Culture', *Keywords for Environmental Studies*. Eds Joni Adamson, William A. Gleason and David N. Pellow. New York: New York University Press, 2015. 50–54. Print.

Rose, Deborah Bird 'Introduction: Writing in the Anthropocene', *Australian Humanities Review* 49 (2009). 87. Print.

Rose, Deborah Bird, T. van Dooren, M. Chrulew, S. Cooke, M. Kearnes and E. O'Gorman. 'Thinking through the Environment, Unsettling the Humanities', *Environmental Humanities* 1 (2012): 1–5. Print.

Snow, Charles Piercy. *The Two Cultures*. London: Cambridge University Press, 2001 [1959]. Print.

Sörlin, Sverker, 'Transforming the Humanities: The Shallow and the Deep, Long-Range Environmental Turn in the Human Sciences', Lecture, 'Entering the Anthropocene Conference', University of Sydney, Sydney, Australia. February 2014. Web.

Tedlock, Dennis. *The Spoken Word and the Work of Interpretation*. Philadelphia: University of Pennsylvania Press, 1983. Print.

Thompson, Nainoa. 'On Wayfinding', *Hawaiian Voyaging Traditions*. 10 May 2016. Web.

Tsai, Huei-Min. 'Sustainability of Small Islands in Taiwan', *Sustainable Development for Island Societies: Taiwan and the World*. Eds H.H Hsiao, C.H. Liu, and H.M. Tsai, Asia–Pacific Research Program of Academia Sinica and SARCS Secretariat at National Central University, 2002. 395–408. Print.

World Commission on Environment and Development. *Our Common Future*. Oxford: Oxford University Press, 1987. Print.

World Peoples Conference on the Rights of Mother Earth and Climate Change. Indigenous People's Declaration. Final Conclusions: Working Group 7. 30 April 2010. Web.

Zalasiewicz, Jan, Mark Williams and Colin N. Waters. 'Anthropocene', *Keywords for Environmental Studies*. Eds Joni Adamson, William A. Gleason and David N. Pellow. New York: New York University Press, 2015. 14–16. Print.

2 Backbone

Holding up our future

Linda Hogan (Chicaza)

Old Inuit Song

I think over again
My small adventures
My fears
Those small ones that seemed so big
For all the vital things
I had to get and to reach
And yet there is only one great thing,
To live and see the great day that dawns
And the light that fills the world.

Distant eagle

For each day that dawns, a new light fills the world. As this old song says, the new light is greater than all one human experience, than all human efforts and fears, even the adventures that once were challenges for the body or mind. This singer understood, long ago, that the world, the universe, is greater than any mere human being in all our strivings.

This is the greatness of an unknown, immense universe in which even this earth is only a small planet in the tail of the Milky Way.

A multitude of languages, human and other, and numerous ecosystems exist all around us. On our living planet these languages may be those chemical and hormonal messages of trees or the communications of animals or insects. There are many precious forms of life, and their ways of knowing go far beyond the intelligence of our own singular human species. These are rarely considered. Yet just as seldom regarded are the knowledge systems and Native science of Indigenous peoples who live in the midst of Euro-Americans or other non-Indigenous peoples. As a Native woman, I am aware that Indigenous knowledge, on all continents, remains largely unknown.

As this world goes through great changes that we have created through human action, it is a necessary, urgent time to enrich and enlarge the content of what we all know of our earth and our own human being here. It is time to look towards Indigenous methods of sustaining ecosystems. Indigenous knowledges are only barely coming to the surface of scientific thought. Many concepts and intelligences I consider immensely practical and viable need to be acknowledged as valuable and useful practices compared to the ones that have dominated our lives for too long.

Some of these concepts are not easily expressed or understood in English. In 'Country and the Gift' (this volume), Deborah Bird Rose introduces one word from Australian Aboriginal thought, the Ngarinman term *Punyu*. This one word holds many meanings. One of these has to do with our cognisance of a sentient world and the world's returned awareness of us. This word holds close a way of life that is an internal connection of thoughts and concepts that altogether keep a world whole and in rightful balance.

This word interests me because it is close to the kind of knowledge system held by Indigenous peoples on all continents. I find it fascinating that we original people are so bound together in our feelings for the earth and for a right way of being, for rules about how to treat the ecosystems and earth.

On this North American continent, one word found in my own Chicaza language is *tish*. This word and its meanings are similar in concept to *punyu*. But before addressing the true meaning of *tish*, we need to consider its contemporary usage and why changes in its meaning have taken place. Over the many years of colonisation and religious forces, the word *tish* was transformed from its original significant meaning to become merely part of another word, *Tishomingo*, now the name of the small Oklahoma town I once inhabited. Placed together with the word 'Mingo,' which means chief, the word has become the name for a single man in our history, known only as *Peace Chief*. He was one of our leaders who either arrived in Indian Territory during the forced Removal of our people from east of the Mississippi River or who died of smallpox along the way. However, the combined two words once represented the position of *any* man or woman whose work it was to keep peace and to negotiate for the people.

Taking those two words apart, the word *tish*, by itself, like *punyu*, refers to a special kind of balance. Its first connotation has to do with *medicine*, not simply as the notion of healing, but as a harmony that embraces how we humans live with all the rest, all the other animals, plants and the elements, including water, which has such vital importance to us. Its numerous layers of meaning not only allow us to see our own small human place in the universe, but carry our original instructions from creation, that our work is to care for the lives that have also cared for us.

Our past villages were on holy sites, lands with our stories, long histories and surrounded by rich food sources. While we were considered to be water people in our ways and location, on the other side of our homes were great forests, and we were first and foremost the keepers of this old growth richness. As caretakers of ancient forests for many millennia, we needed that balance and vision and the foods that were there. We held extraordinary knowledge relating to trees, including the certain dappled light that forests need for understory plants, air currents travelling through leafed branches of a tree and the incredible wealth of soils and nutrients. In locations such as Cahokia, with human populations larger than those in London or Paris at the time, it was important that we understood a workable sustainability.[1]

Our forests were cared for so well that the invaders noticed and wrote about them in the 1500s, when Hernando de Soto, the Castilian conquistador, and 700 of his men entered our world. Their journals, in Spanish collections, speak of the beauty and functionality of the forests and fields that were like nothing the Europeans had seen. Our care for them was impressive. It was our work to keep *tish*, health and harmony, not only in the forests, but in our villages, agricultural sites and beloved medicine gardens. Botanists, travelling with each group of intruders, wrote copiously in their notebooks about the forests and the fields, even the berries, and included sketches of the immense agricultural regions and cornfields stretching beyond vision.

Beyond vision may best depict what I want to say. It also has to do with the concepts of *tish* and *punyu*. These kinds of concepts exist in all of the Native knowledge systems and languages I have studied or heard spoken by various tribal peoples. The most well-known of these is from the Navajo of the American Southwest. The word is *Hozho*. In order to maintain *hozho*, or balance, sometimes extremely elaborate ceremonies are performed. These are looked over by specialists who may know only one song, one long prayer or part of the ceremony. Most of these ceremonies continue today, using complex myths, songs, prayers and actions. As one priest said, their complexity, and the fact that they continue to be remembered and performed, is proof of their validity.

These ceremonies allow for a human or community to know their place within a circle of life around them, animals, plants, the earth, all extending into the universe. It rights a people, or a person, to their place within that great day dawning, a cosmic environment, and into a circle of kinship with all living creations.

In Northeast North America, the Haudenausonee Thanksgiving Address given at the beginning of gatherings also brings the mind of humans into a state of conscious gratitude for their surroundings, thanking even the air people breathe. It, too, is a placement of the people with their surroundings.

Tish, punyu, hozho, all words having to do with vision and balance, allowed for millennia of sustainability. It is still possible for people in the present time to look back to these concepts and carry this knowledge into the future. These concepts from around the world exist in Indigenous knowledge systems and languages, related to the totality of survival that includes the invisible realm of stories and events that have long been a part of certain lands.

If these notions of balance had been adopted by the conquering British on the east side of North America, or by the Spanish and missionaries who invaded and enslaved the west coast, we would live in a different world today. Had there been a cross-fertilisation of ideas and knowledge, we would not be having discussions about the suffering and corruption of the planet. The term 'Anthropocene,' a problem in its own ways, would not be a part of this current conversation. Damage to sacred lands may have been averted. Each form of life would have been considered as integral to the whole, and co-existence with other species would have been a natural extension of our own. However, Indigenous knowledge was considered inferior, and it continues to be. Instead, a destructive sense of superiority and focus on individualism still dominates all our continents.

After the arrival of the invaders to our Mississippian Southern homelands, European industries felled the great forests to fulfil their own needs. Few people realise that a European corporation, The Mississippi Company, later the Company of the Indies (1684–1720), was created and run by murderers and gamblers, but recognised as legitimate by France. They held a business monopoly in the French colonies and West Indies and caused one of the earliest examples of an 'economic bubble'. The 'Mississippi Bubble' burst in 1720, after exaggeration of wealth and holdings in Louisiana associated with wars and the enslavement of our people in island sugar economies, led to wild speculation on company shares.

An economic 'bubble' is a term some of us think of as new, but even in the 1700s, economic speculations were creating environmental destruction. Economic notions of wealth have only increased in scale since that time and are a root cause of our problems today, as many continue to live amidst deforestation, mining and the ongoing destruction of water and land, and sometimes even as slaves. Warfare still breaks entire ecosystems and any human survivors.

Watching this destruction, Native Peoples are left feeling that what we know makes little difference in the face of the desires of others. Decision- and policy-makers place personal considerations over the survival of people and land.

We can calculate the damage into numbers and describe it in scientific writings for our politicians to read, but the corporations will continue to tempt powerful people. The question I ask now is how do we create change? Or is it implausible to expect great changes in a time being described in the hope-breaking language of the *Anthropocene*, a word I would like to disappear from our conversations, except when it is being used in its original geological contexts.

In considering how to create change, story is a significant tool, and the word has power. Politicians know this. Preachers count on it. It is a valuable force used for changing thought. Stories have been our creators. It is through story that we know ourselves. It is through our stories, potent as the narratives of the sciences, that we Indigenous peoples learned about our environments and our ways of living within them. Stories are the backbone of all our life-ways and spirituality.

As a younger Native writer, I began thinking of how language could be a form of change. The only true gift I had to offer was the use of words and stories. If I could tell the truth about how people are affected by environmental change, perhaps it might touch the hearts of others and offer a means of transformation. I hoped that if readers could see and feel the stories of others, it would be a path to different ways of living, a new narrative for the direction our lives might travel.

I looked to the stories of our world, our cultures, our selves. In stories of the present time, I think that if we could strongly use words to touch the imaginations of others, they might see the trees falling as the permafrost melts, and animals losing their environments, or tribal people having to move their hospitals, schools and villages away from the threats of rising oceans. These events might be seen, imagined, and grasped more fully in their magnitude.

In my own work, I used the Hydroquebec dam project in *Solar Storms* as an example of this verbal vision. This novel fictionalised the true story of an enormous dam that was being built in Quebec, and the corporation creating a grid to sell electricity all the way to Texas. When New York discovered how much damage was being done, they cancelled their own contract with the corporation.

In *Power*, another fictionalised story of true events, the main event is the killing of an endangered species. These writings have made the difference I hoped for among those who read them. In classrooms they continue to change the consciousness of others. They reveal the devastated and toxic environments of the Everglades and the torn Northern communities devastated by Hydroquebec dams. With *Mean Spirit*, I wrote about the greed and corruption of the first oil boom in Oklahoma, thinking it might bring to light not only the plight of Indian nations that were affected, but also what oil wealth means even now to people in the middle of its drilling and production. Years later, I lived with earthquakes created by similar violence to the earth, with the oil ponds from fracking, watching enough gas burn off from the rigs to heat all the small towns near Tishomingo.

With ideas of connective creation, we have been putting this world back together, not only in creative form, but also in some of the restorations that are taking place.

For example, the people in far North Cree country have taken back the original names of their lands. The place names unite them with creation and creator. The villages are given names such as The Creator's Elbow, The Creator's Knee or The Creator's Eyebrow until earth and body equals first creation. This arrangement, with The Backbone of Creation, places the world in a different visual, spiritual, and cognitive order. It also aligns the earth world with tribal astronomy.

This kind of re-naming is something akin to *tish* or *punyu*; it is a manner of keeping the world in balance, a daily reminder that the land is sacred and to be cared for as the living body of all that exists.

In one of her newer works called 'Backbone,' First Nations dancer and choreographer Sandra LaRonde also recognises that humans are inseparable from the earth. About her newer programme, she says,

> The idea that there is a spine to our continent is an Indigenous concept. It reveals our way of perceiving land as sentient, alive, and intact. I wanted to show the 'backbone' of this continent in dance and music, and that this rocky mountainous spine has life, circuitry, electricity, and impulses that are active and dynamic – much like the human spine. For Indigenous peoples, there is not much difference between the earth's backbone and a human one. We are inseparable.
>
> (n. pag.)[2]

With her creative work, LaRonde expresses the way we are part of the body of the earth in many realms including that of bodily, physical electricity. The human and the earth body have the same cartography, movement and life energy. As in naming the land after the creator, the beauty of the world is the equal to the human place within it. This is *tish*, *punyu*, *hozho*. The dance functions as contemporary ceremony and shows us that we are earth. We are universe.

In this same connection, as a daily reminder we step out in the morning to pray, or go to the water with our tobacco, smudge our home. This is part of a daily ritual that keeps us as one with the world, sharply aware of the earth body. Our words are spoken to, for and with the earth in an exchange. Like the birds that pass over, the animals in the distance, all hears us, knows us and sees us. Central to our thinking as traditional Indigenous peoples is this relationship. We dance, sing or write all that is above and beneath. Sky. Earth. Water. Rock.

I understand that in Aboriginal law, the same entwining of human with land, water and the universe takes place in the present as it did in the original creation. When Deborah Bird Rose brings into this conversation the Ngarinman term *punyu*, she also states that the land is alive and aware of us; it is part of us and listens to those who speak. *Punyu* is pre-historical in origins. On the Australian continent it begins with the Dreaming, original law, and this knowledge that all beings are sentient, including those of the past. We are all participants in the life of the world and know that the creators, the first people, the First Ones, remain present. They continue to speak and work within and through the present, as our ancestral cell system, blood and breath. That this Indigenous understanding of the world extends around the globe is important. Most first people still work to keep their lands alive and viable.

Returning to mapping land and body, Deborah addresses cartography in terms of the Australian continent, the first Dreaming that exists as a spiritual map of the earth far different than the boundary lines of European cartographies that have to do with ownership, political boundaries, or land purchases of the past. Deborah has said that for the original people of Australia, sacred and spiritual terrain where events and stories of what has taken place in the past and in each landform are mapped. These maps survive whole and healthy into the present where life and remembering is ongoing and still is in process. Human beings are still acting upon the land, caring for it. We are responsible not only for recalling historical stories of importance but for maintaining sustainability towards the future. In addition, our work is to remember the original attributes of the land. The notion of a spiritual map is important to hold close. So is the knowledge that we are still a world in the process of creation, a dance, from a long ago beginning that moves into the future.

For all Indigenous thinkers, Natural Law takes precedence over the creation of human laws. We are responsible for caretaking of land and not allowing disturbances or destruction. We are thinkers for the future.

While I've been told this is not the best example, what also comes to mind is the 1984 film by Werner Herzog, 'Where the Green Ants Dream,' when miners could not comprehend why the Aboriginal Miliritbi owners of a region in

Australia could not permit the mining of a holy site and had to resist miners who wanted to blast the land for uranium.

For our Indigenous thought-makers and dreamers, life and time in the world moves in cyclic events, like seasons, and not as it appears in the storied realms of Western thought where the world is falling away for good. This 'for good' is an oft-used term that requires close scrutiny. That the end is for good has interesting connotations. Towards this end is an embedded belief that the prediction will not only come to pass but that it is in some way 'good'.

In a linear thought system, the end is coming. In Biblical stories, everything moves towards this apocalyptic vision, the time of Revelations. Life travels in a straight line until it falls over an edge. Many religious people await these times because they expect a better life than one they are living on earth. The move towards heaven, or 'the rapture,' are both mental constructs based on a belief. This 'end' appears as a constant in European literatures, with its history of warfare and a doomed future. For many people, their thoughts have been so steeped in this ending that they are no longer aware that this thought is practically carried in the cells of their bodies.

This consideration is separate from those who have learned through preceding millennia that the world they care for will exist for all time and future generations. It makes sense to contemplate and deeply reflect on the creative stories and works of the Western view, from children's stories to every word that goes into the minds and hearts of a people. Comparing these views and understandings, a cross-fertilisation of ideas, as I suggested above, is important if we are going to have a survivable world.

As Deborah also says, the entire universe is the expression of the First Ancestors who created this world, and then lived inside their own creation. Living inside the creation specifies that the ancestral beings remain with us in a world that is still a living, thinking and breathing organism.

We are holding the future. Its rocky spine rests across our lands and we must be careful with what we make of our environments. This relatively new term, the Anthropocene, concerns me in a number of ways. Foremost is that it holds a story which may come to the fulfilment of its expectations. It is what I mean when I speak of the end as part of a way of thinking about the world now and in the future.

The Anthropocene ignores those people at work in the material world, humans doing labour or fieldwork that is creative and restorative, rather than that which is destructive. Many of us long for a niche where we can help change the process of our ongoing creation, transform the trajectory of the changes destructive enough to give us such terms as 'The Sixth Extinction' and 'The Tipping Point'.

A single world like the Anthropocene can forget, even disappear, long periods of sustainability that have existed and can exist again. On our continents, we went from understandings, concepts and ways that kept a world alive and growing despite large populations. Words such as Anthropocene take in and consider only a brief period of destruction.

History tells us that great destruction had already taken place on the European continent before Europeans arrived here to occupy lands they considered 'new' despite the ancient histories of the people already living here with their own histories of thriving environments and successful production of foods. We must ask if they ever examined the reasons for their own failures. Did they never reflect on the reasons for the loss of their crops, the damages to their forests or the lack of clean water? Why now do we live in their story? It has been one of continued slavery, deforestation and wars that break entire ecosystems and their peoples.

Just as I have so often wondered why some have not learned the notions of balance or care, *tish* or *punyu*, I also wonder about the lack of love for this world. It seems that both the failure of mind *and* heart arrived across the oceans. This colonised failure still rarely considers Indigenous practices of caring for the land. The very same practices put into place on the 'new' American continents against the will and the knowledge of the first people present were then also put into place on other colonised continents.

Our ways of thought need to be re-examined at the core and then revised. Consider the Greek tale of Pandora's Box and how all spirits were let loose except for 'Hope' who remains locked inside. This is fortunately only one container of this idea, one story. It is a myth from only one way of knowing a world, but then doesn't myth itself contain a civilisation's deepest thoughts? If these ways of knowing begin in childhood, think of the numerous stories and mythologies that become the thought-ways and cultures of the colonising people, especially when Western myth or story rarely shows a way to ongoing survival, offers hope, or even happiness. They are filled with people who are unkind, deceptive or who threaten other species. The majority of these stories, with all hope locked away, do not tell of continuance, or a world ongoing. They do not teach that animals or plants are companions. Yet these stories are implanted into the imaginations of the dominating cultures.

Let us suppose there is another kind of box and it is fully emptied. Or let's change stories altogether. Hope is not held back, but also escapes. In this story, hope exists as a part of our world. And hope asks us to act, to consider a future. It requires that we create a change of mind with thoughts for the future and the entire community, including the non-human. This is what the Iroquois mean when they say we must make our decisions always considering seven generations into the future.

The use of story is a strong part of our lives and a powerful tool in religions. In the Christian creation story a singular god created a beautiful and lush gardens in the place where rivers met. The problem, as with Pandora's Box, is the challenge given the first people in this garden named Eden. Although they are loved, no other first people in any Indigenous creation story are challenged in such a way as Adam and Eve, tempted to eat of a divine food that will offer them knowledge. Any intelligent god would know that 'his' children would consume that fruit, would one day come from innocence into intelligence and knowing.

Instead of learning their relationship to this world through the words of their creator, instead of the instructions that will keep a world alive, Adam and Eve are simply sent away. They are removed from the natural world, even from the divine by the simple act of seeking knowledge. Most of us know this story. In a way this is the story of the end of stories in this particular culture. After the shunning from the garden, what follows are stories of destruction and war, deception, betrayal and lies. Where are the stories about relationship and great love, or as with our own about how a canoe with singing women came down from the sky and one of the young women fell in love with an earth man who returned the love and their struggle and success in overcoming differences. Where in Christianity is the humour? The creator is not similar to humans who have flaws. It is as if that first sin was the ending. There is no longer a good garden, no forest. The people do not know the story of their land or world, or learn that they are one with the spine, heart and breath of all the rest. They are not a part of it.

In most other creation stories, the animals help the humans when they are created or when they come out from the earth or down from the sky. The animals and plants have always helped our species and they are all loved and cared for greatly as kindred. The stories of our creation continue across continents and the words enter our human minds, spirits and hearts in many ways, influencing our lives and treatment of this world and its beings. Like song lines that cross a continent and map past events, language and story is the map, the anatomy and the living backbone of our cultures.

Still, we have to consider that we now live inside something incompletely mapped. The lines have been changed. The stories are not ours. We are not at an end, and for us we live inside a story still in its creation.

For most of us, humanity is connected with the land, and with 'caring for country,' as in the true sense, as being woven into one inseparable living thing, most first peoples understand the proper, sacred way to relate to the ecosystem in which they live.

After hearing Deborah speak of the many meanings and uses of the word 'country' in Australia, I researched the term 'Caring for Country' to be certain I understood. It fits with the same concepts of *Tish*, *Hozho* and *Punyu*. In their original meanings, the other three words include the strengthening of cultural and environmental bonds. These words speak to all our unique places within the world. The words there also include community of all kinds. It may be wilderness or bush, but it includes the human as well as all other forms of life. It contains the invisible pathways overcrossing one another on our lands that are related with the ancient ones. Deborah may correct me, but I think land holds within it all the songs, prayers, and ceremonies, the histories and stories, just like those of our own forests or waters. It is a term not of separation but one of community in its largest sense.

When Aboriginal people use 'Caring for Country' it is a way of saying all are intertwined, that there is a braided connection with the land, its past, its ancestral histories and all the stories present in this great assemblage. This is a far distance from the view of governments when 'country' and 'caring for country' takes on the meaning of ownership or management, as has happened, as the term is often used with an eye only to one problem at hand and not seeing the whole. In Indigenous terms these words do not have to do with land or forest management, changing it, or even as Deborah states, adding toxins and pesticides to create a place that fits with the human plan, not truly a Caring For Country. The human plan alone does not fit within that of the country in all its numerous dimensions and connections.

'I love this land I walk on,' says Warren Clark, at Mungo National Park. Warren is chief Aboriginal Executive Officer of the Mungo Joint Management Advisory Committee. He says, 'Coming to Mungo I get a different sense of feeling, that I'm home. You seem to know when you're back in your own Country. It's not taught to you, it's built in you. It's in your soul. That, that's your Country.'

If others could learn to love the land this much, it might stop the advance of the Anthropocene. With all our emotional pain on the one hand, the sense of superiority on the other, we may learn to love and to keep with the spirit Mr Clark elucidates, the one he calls the soul. This feeling of home has kept us turned towards our older homelands and traditions, and respectful of other Indigenous peoples' deep knowledge. It is built into the human, a part of the body, again a backbone which is part of the land.

Learning to love the land and know the land, the water, the air, all as our spiritual terrain doesn't mean we can forget or ignore the errors of history. And again, we have to look at knowledge systems that have held up the world for much longer than those of the past few broken centuries. It has been a long journey of research for me to discover some of what allowed Indigenous peoples to survive. A vast abyss lies between different understandings of being that exist in diverse knowledge systems, and we must become more aware of the ways Indigenous and Aboriginal peoples managed to survive and to prevent so many lives from transpiring.

Within a new framework for thinking, growing, and for teaching, there is room for hope. We have to hold to this spirit of possibility. *Hope* is the essential and necessary word that gives us our human direction. The Anthropocene, as a term, is a closed container. The lid is tight and, like Pandora's Box, it fails to allow for any other ideas or changes, therefore hindering any positive future outcome for all creation.

It is primarily scholars who use the word 'Anthropocene' to point to the violence being done to the planet. Others are not only aware of the violence but have experienced it. Environmental injustices affect those in third world circumstances more than others. In my life, I have met many people who have long been trying to find the road to walk away from this destruction, and this includes scholars and artist activists as well as grassroots organisers. These people are changing the story. They have united in restoring the world, returning ancient seeds, reigniting older Indigenous knowledge, even growing foods once thought extinct. One of my relatives created The Chaac Project to clean water for people in Mexico by re-routing the water flow and planting trees, all in a location where the people had been dying of cholera and other diseases carried by contamination.

Other recent events help us greatly with understanding the term 'caring'. For example, several years ago when Alex White Plume restored the bison to Pine Ridge, everything in that region began to change after the buffalo were released.[3] The weight of their hooves and massive bodies brought water to the surface in streams, springs and creeks that had not been present for many years except in old stories began running again. The elders remarked on these returned waters and the changes they were seeing. Soon grasslands with original grasses came back, as did insects that once lived among these grasses and waters. The insects brought about the return of birds. Soon other forms of wildlife returned and the land began returning to what it had been in the past.

When Wangeri Maathai began the Green Belt Movement in Africa, she didn't know her work would result in a Nobel Peace Prize-winning project. She and her people began planting many trees. With the continued planting of many trees, water began to be restored to the once parched, dried and cracked land. With that water, and with the new green growth, the animals returned. So did the birds. Many food sources also grew once again. Other kinds of plants rose up in the undergrowth, including the old medicinal plants. The planting continues with those who assisted Maathai and then taught others.

The soil, cared for, is capable of growing what we plant. The bacteria beneath surface earth helps trees send chemical messages to others when dangerous insects or people approach. All living parts of the planet work together if we will only permit it. This allows for balance, *hozho*, to return, as bison may have done, as new forests do. It is only a small part of the concept of *tish*.

These few, seemingly small forms of restoration, the bison, the planting of trees, have created great changes, revealing to us that there are ways to restore our lands to something that does not leave us at a 'tipping point,' as Bill McKibben says so often about where we now dwell. The planting of trees not only provides water and food, but also sequesters carbon and airborne chemicals, and planting mangroves desalinates ocean water and creates savannah, which may eventually become grasslands or forest.

While McKibben is right in regards to climate change, the notion of a tipping point takes away hope. Without that hope, too many give up and prepare to face a bleak future. I do not want to pretend it is not bleak; I would never have imagined a time when German companies charge poor Africans for a small bucket of clean water. Who would have thought water could be privatised and owned, ruining the lives of those who cannot buy it. I can't overlook fracking when I have experienced the earthquakes and oil fields near my home in Oklahoma. And, like others, with despair, I have watched species disappear. Those who deny truth are largely the ones whose incomes increase with contamination and violations of nature. Flint, Michigan is an example of that. Those who worked in government knew the water was polluted with lead and other toxins and used clean, uncontaminated water for their own families, while allowing others to become victims of poisoning.

But the examples of action reveal to us that there are possibilities and ways to pull ourselves out of the great sinkhole of damage that has been created and I know that we can bring together our different and unique ways of knowing to find possibilities for a better future. We live with responsive lands and a responsive earth. It welcomes and desires healing and restoration. The earth is a receptive life force.

I have to wonder if it is even ethical to discuss one word, as we are now doing with Anthropocene, while so many are working towards restoration and towards returning what was taken from the long ago sustainable pasts of Indigenous peoples. From preservation of native seeds to the caring for full-grown trees, animal life and grasslands, restoration is taking place everywhere. The returns are numerous. Plants are reintroduced that will save certain necessary pollinators or butterflies. Others will be nitrogen-fixing for planting food in the future. Serious work is being undertaken by people who have learned their ecosystems intimately and are able to participate with the healthy interactions and connections.

The rest of us need to search out all these situations and our possibilities and put ourselves to work.

It is time for us to quit haggling over single words or concepts, to give up the vision of the singular human self, change our education systems and get to work ending the threats to our environments.

The word 'anthropocene' is useful on its linguistic surface, but it simply does not reach back far enough to take in our Indigenous possibilities for sustainability. It looks only at the one Western system of thought and being. In that, it seems to be a self-centred word that focuses only on human lives, and in that it becomes an arrogant model. It has become a template into which everything literally falls, every form of life, every map of being, and one that leaves no room for any discussion, change or hope.

We have interfered with rivers, animal migrations and other unwanted changes to the planet, but we have to avoid the sense of hopelessness and the accompanying depression and inaction that accompanies it. We need to consider the question, "Why can't we stop?"

Any answer to the question requires both a political and a cultural change that needs to take place now. As Deborah writes in her chapter for this book, we need a true 'Caring for Country'. We know what actions are not the right ones. Our spiritual core recognises Natural Law and we have a requirement and responsibility to act with integrity, to care for a world in a spiritual manner.

Something deeper than just a 'shift' has to take place in our time. It may come from the Backbone of human spirit and its felt connection to the earth. We are small beings but as in the opening song quoted above, we want to pass through our life of growth to a light-filled day where we know our world and the lives remaining here will be cared for. We can't afford apathy. We must prepare to enter the full and sometimes difficult process of Caring for Country. What is left in the wake of past destruction remains to be transformed into a harmonious and full, global bid for life.

Notes

1 Cahokia (c. 600–1400 CE) is a pre-Columbian, Mississippian-culture city built in the Missouri River watershed in North America and estimated to have been populated by up to 40,000 people.

2 See 'World Premier of "Backbone"', www.banffcentre.ca/articles/world-premiere-backbone (17 August 2015).
3 Pine Ridge references the Oglala Lakota reservation in the US state of South Dakota.

Works cited

Hogan, Linda. *Power*. New York, NY: W. W. Norton & Company. 1998. Print.
——*Mean Spirit*. New York, NY: Atheneum Books, 1990.
——*Solar Storms*. New York, NY: Scribner Press. 1995. Print.
McKibben, Bill. 'The Tipping Point'. *environment360*. 3 June 2008. Web.

3 Country and the gift

Deborah Bird Rose

In 'Backbone: Holding Up Our Future,' Linda Hogan (Chicaza) addresses questions that hit hard in this era we call the Anthropocene, and one of her great forms for conveying understanding is 'backbone' (Hogan, this volume). She signals the fleshy, embodied, living and structured quality of Earth life. This is not abstract; there is no determination that metaphor takes precedence over living reality, and, if I understand correctly, the Western distinction between organic and inorganic matter is irrelevant within connectivities that permeate all matter. Backbone is part of life across numerous scales and types.

At the very outset of this conversation I want to acknowledge the Elders who have taught me. I could not understand Earth life as I do without having sat at the feet of Indigenous Elders in Australia, in North America and in Taiwan. In addition, my thinking aims for connectivity and porosity, and I have learned a great deal from Western philosophers and other thinkers.

My words address the fact that the Australian Aboriginal practice and 'Law' of 'Caring for Country' form the basis of a conversation we need to be having. It is a conversation for the Anthropocene, which is to say that it is absolutely urgent. The urgency can be better appreciated if we translate Anthropocene into more vivid and clearer words: 'in the midst of great and growing destruction'. In pursuit of conversations toward these urgent issues of our day, I aim for encounter, not division. There is a non-negotiable stratum: Earth life. My deep desire is to hold fast to its fleshy, embodied materiality.

The conversations we need to be having in the midst of great and growing destruction are deeply challenging, and there are real difficulties when we refuse to homogenise the differing histories and ontologies that are party to heterogeneous dialogue. Multiple species are part of this conversation as well; the issues that confront us are not only about us humans. Finally, by way of introduction, I need to mention this slippery word 'we' which might include every living and non-living thing, or could include only animals, or could include only humans, or might be understood as including Western culture but not others, or indeed there may be many other possible ways of drawing boundaries and forming inclusivities, each of which has a reasonable place in conversations concerning the Anthropocene. It is interesting, therefore, that the Latin roots of the word 'discourse' have to do with running around all over the place. The word 'we' absolutely does this: I use it discursively, it runs to and fro in a most unrestrained manner.

In my book *Dingo Makes Us Human* I wrote that 'anthropology is, or so one hopes, a thorn: a discomfort to those who like their worlds and ideas neatly packaged' (*Dingo Makes Us Human* 236). Our conversations for the Anthropocene must be like this. They must be disruptors. In fact I hope they will be large thorns, and that they will lodge themselves in sensitive places. Linda writes about the fact that we are needing something more than a shift in attitude. Large uncomfortable conversations are a good start. And actually, this is what we in the humanities often do quite well – we shake things up. We offer critique. We make things difficult, and in the environmental humanities we make things difficult in relation to environmental thought and practice. I fear that we have not yet learned to be anywhere near difficult enough.

The magnitude of the events that coalesce into the process we label 'Anthropocene' are in fact unimaginable. We know the climate change issue well because it has the greatest profile, but it is just one big part of a much wider set of entwined events that include the great mass extinction event now in process, the acidification of the oceans, the accumulation of plastic waste, the loss of soils and fertility, the industrialisation of death in many forms from factory farms to pesticides and herbicides, the loss of rainforests and the rampant consumption that fuels the work of tearing up and wrecking planet Earth. And let us remember the cruel twist of modernity: much of this wreckage has been and is now being done in the name of human progress – that is, to secure (an allegedly) happier, more liberated and healthier humanity. Linda's chapter is an exquisite reminder that the term 'Anthropocene' fails even to do justice to the experience of the human species which would seem to be at the core of the concept.

Country

Aboriginal people in Australia have picked up the English word 'country' and used it discursively; they have absolutely run away with it. In fact, the Aboriginal use of the term has been so inspiring that it is quietly being picked up and moved back into mainstream Australian English with a new resonance absorbed from its life as an Aboriginal term. The fact that the commonly used phrase 'caring for country' does not include a definite article is indicative of the some of the shifts that are taking place in mainstream English as a result of conversations with Aboriginal people.

I am thinking of Country as a being (not a 'thing') and 'Caring for Country' as a practice (not a 'policy'). In Aboriginal English Country is a subject. Country is multi-dimensional: it consists of people, animals, plants, Dreamings, the dead and the yet to be born, underground, earth, soils, minerals and waters, surface water and air. There is sea Country and land Country; sky Country too. Country has origins and a future; it exists both in and through time. Country is the place of creation and, as Linda tells us, that creation is on-going. The sites of Dreamings' or creators' actions are in Country, and their tracks criss-cross Countries, connecting one to another through the great songlines or travels that were at the origin, and now are at the centre, of the on-going-ness.

Country cannot be defined in an exhaustive sense. Like a person, for example, life and will are on-going. We do not know exactly what will happen next, what Country will do next, what life will do next. Country has its own story, its people and everything else are part of it, and from generation to generation, that story is not over.

In exactly the same way, 'care' is not a thing. It is a practice, a multispecies, multi-temporal practice, and it is effective insofar as it is adaptive and flexible. Care is reciprocal. As my friend and teacher April Bright explained: 'if you don't look after country, country won't look after you' (Rose *Nourishing Terrains* 61). The story of care is never singular, and nor is it ever settled.

In sum, there are many parameters to both 'caring' and 'Country,' and that means that there are many ways into the conversations we are having as we seek to address the urgency and magnitude of the great destruction. I will explore several paths, focusing on those that seem to offer particularly interesting difficulties.

It will be helpful, though, to begin on a cautionary note. As the phrase 'caring for country' has been moved into mainstream English it has frequently been domesticated, de-fanged and de-clawed; its capacity to provoke genuine change has been appallingly neutralised. This is a common and widespread occurrence: the practice is to take an inspiring idea out of Indigenous culture, re-purpose it in mainstream culture and, all too often, end up with something that trivialises, or even totally destroys, the meanings which had seemed inspiring when they were with Indigenous people.

A brief example will elucidate. At the Australian government's NRM (natural resource management) website one reads:

> *Caring for our Country* [please note the NRM's insertion of the possessive pronoun 'our'] is one way the Australian Government funds environmental management of our natural resources by supporting communities, farmers and other land managers to protect Australia's natural environment and sustainability.
>
> Funding Indigenous projects is an important part of Caring for our Country. It contributes to the Council of Australian Government's commitment to Closing the Gap.[1]

There are numerous oddities and elisions in this statement. For a start, I am struck with how agency is distributed and effaced in the first paragraph. The programme is the active agent, and it works to help a number of humans in their efforts to 'protect' the environment. This inversion of agency effaces the power of Country. The ground of all life – Country – is posited as the passive beneficiary of a government programme. In the second paragraph Aboriginal people are brought in, apparently as a secondary category of beneficiaries, after farmers et al. In addition, it seems that the beneficiary of the program is actually another set of government programmes, those concerned with Aboriginal people. In short, a beautiful term is borrowed, transformed and deployed in a way that appears to eradicate the power and active presence of both Aboriginal people and Country.

Worse follows. This programme includes the National Landcare Program (the term 'landcare' seems clearly to reference Aboriginal thought).[2] A publication from Landcare Victoria is called 'Landcare Notes: 1080 Poison Baits for Pest Animal Control'. This set of information is based on the assumption that good land care involves using deadly poison to kill animals that get in the way of primary producers. The info-sheet discusses the merits and lethal doses of 1080 (*sodium fluoroacetate*, a poison that causes terrible deaths and is used particularly, but not exclusively, to target canines). As the project unfolds, we see the process of appropriating an idea that was inspiring (Caring for Country), distorting it to suit mainstream control (Caring for *our* Country), linking it into capitalist systems of production and death, routinising and standardising it as policy, and presenting it to the public as best practice in relation to that beautiful, and now totally inverted, aspiration to care for Country. Spreading poison, which is to say spreading suffering and death, all over the land is more than a failure to understand caring for Country. It is a heartbreaking leap from mimicry to mockery, by way of industrialised killing.

This brief example, one among an alarming number that could be brought forward, shows a determination to inhabit the land through all the old colonising techniques: dispossession, appropriation, replacement and slaughter. The example shows some of the reasons why we need these difficult conversations.

Punyu

Country is a living entity, the on-going interactions of many living entities including some, like water, that Western thought generally classes as non-living. When it is in good health, Country is *punyu*. This Ngarinman term has cognates in other Aboriginal languages of Australia, and Linda connects the term with other tribal peoples. At the heart is its capacity to hold a multitude of creatures, with their great and varied desires and practices, in one encompassing (but not homogenising) matrix. It is, perhaps, a backbone.

There is no a single English equivalent for this term. *Punyu* is all that is good: *punyu* means healthy, good, beautiful, knowledgeable, happy, responsible, clean and within the law (that is, living properly). This is a recursive mode of being and becoming, it constantly feeds into itself, generating more of that which is good. For humans, *punyu* is a state of maintaining one's own health, promoting that of others and promoting the health of Country. The term does not distinguish between physical, mental, social and spiritual being, nor does it distinguish between humans and other living creatures. Country is included within the concept of *punyu* – Country too can be healthy or sick, clean or uncared for. Country can be *punyu*, or it can be broken down – *wanggut* in Ngarinman.

Alive in Law

Another great conversational path is 'Law'. The Aboriginal philosopher Mary Graham identifies two basic precepts. Each one is deceptively simple and extraordinarily complex.

- The First: The Land is the Law.
- The Second: You are not alone in the world.

(Graham 35)

Each of these precepts is incredibly challenging for contemporary Western thought. Stephen Muecke has offered the absolute best meta-statement about Aboriginal philosophy, with his characteristic humility: 'If I were to risk a generalization I would say Aboriginal philosophies, as ways of life, are all about keeping things alive in their place' (Muecke 34).

What might it mean to say that 'the Land is the Law'? Ambelin Kwaymullina and Blaze Kwaymullina present an elegant analysis in their excellent article 'Learning to read the signs: law in an Indigenous reality'. Like Mary Graham, they emphasise relationships and relationality. They are interested in contrasts between mainstream Western thought and the Aboriginal mode of Law, in which relationships come first. Learning and responsible action are shaped from within relationships. Ambelin and Blaze Kwaymullina tell us that Law is part of all life, and there is nothing inanimate in Country (Kwaymullina and Kwaymullina 200). In their beautiful words: 'Law flows from the living hearts of Aboriginal countries, and in this sense is location specific' (202). They discuss the Nunga lawyer Irene Watson, who writes about the state of being in exile from the Law. She attributes most of the disasters and crises the world is facing to the fact that the greater part of humanity is living in exile from the Law. And actually, I think she is putting it rather kindly. The majority of us, let us be clear, are outlaws. Willingly, or not so willingly, and indeed for many of us not at all willingly, we are participants in this great destruction. We are not working within the Law that flows from the Land.

Our conversation about caring for Country is one of the ways we can start to transform this outlaw condition, for the move from outside the law to inside is also a move from outside of Country to inside. Even as I state this as a conceptual shift, the effects are practical. Country requires work. *Punyu* is achieved by the many participants who are part of the interactive and care-giving Earth life.

Insiders, one and all

The West's nature-culture binary posits a radical disjunction between the human world of culture and all else on Earth, including the biosphere ('nature'). In its political formulation of types of humanity, it further distinguishes those who are highly evolved and who understand themselves to be outside of nature, and those who are less evolved and think of themselves as part of nature. In the so-called evolved form, the idea is that culture is something like a platform from which humans (by definition the only active intelligent agents) act upon a passive and non-agential nature.[3]

The complexities of how the binary works as a trap came to me vividly when I was travelling in the USA with one of the co-authors of *Country of the Heart*, Margaret Daiyi (Rose *Country of the Heart*). We were doing a radio interview with a sympathetic presenter who wanted to know more about the use of fire as part of caring for Country. Every question he asked already presupposed a managerial paradigm in which humans were outside of, and acting upon, Country. Margy was finding these questions difficult to answer, and finally she blurted out: 'You have to understand: we are part of the flora and fauna of our Country.' She got a wry look on her face as she said this. In her mind, as in mine, was the knowledge of the awful history of colonisation during which for generations, most Aboriginal people were not counted in the census, and did not exist as citizens or subjects with the same rights as other Australians. The popular myth that Aboriginal people were counted amongst the flora and fauna of Australia is not true, but fits logically with the exclusion they were suffering, and speaks directly to the fact that given the West's nature-culture binary, it was denigratory and perilous to be classed outside the citizenship sphere of culture. It is not clear, for example, that human rights apply to creatures on the wrong side of the binary. That was certainly the Australian experience. And that is how these Western binaries work: there is the valued side, and there is the disvalued side. When you take away the valued side, the default position is disvalue.

Nevertheless, Western binaries are not the only way of inhabiting Earth life. To be on the inside of Country, on the inside of Law, is, as Margy said, to be part of the flora and fauna. This is not a homogeneous space: there are many parts, including the human part. In Country, flora and fauna are all part of it, and are part of networks of relationships. Moreover, these relationships are not homogeneous. They are networks of kinship. Within the domain of Country and kinship, the nature-culture binary topples. It is not a matter of collapsing 'culture' into 'nature,' or 'nature' into 'culture'. Rather, far more interestingly and provocatively, Earth life is all about kinship.

To live in goodness, to live within the Law, to live a life that is *punyu* is to live within relations of kinship. This is not a metaphor. Taken in all seriousness, kinship poses difficulties for Western thought because it is so personalised. In contrast to Darwinian kinship which is impersonal and in some sense accidental, Indigenous kinship is personal and intentional. We (humans and others) are claimed by our kin and claimed by our Country. That claim is the assertion of a relationship of mutual responsibility. Australian Aboriginal people speak of being born for Country, born from and for the ancestral or Dreaming power of Country. As with the family kinship that is more familiar to Western peoples, a person comes into the world already embedded within relationships of responsibility. Adults take responsibility for the helpless young, and the young grow into adults who (provided all goes well) are able to fulfil responsibilities. When Country is understood in the manner of kinship, it follows that a person is born into relationships of responsibility: Country (in the form of food, water, sun, shade, air and much more) nurtures the person and in due course the person learns to take care of Country.

Within the kindred

Kin have a regard for each other. To live within a world of kinship is to live within responsibilities that are wide-ranging, and that always precede the life of any given creature. I am not the creator of my responsibilities; they come before me, and they will go on after me. They are not unbounded; they are focussed on Country. But Country itself is multifarious, and so my responsibilities include a multitude of others. I have emphasised Country because my focus in this discussion is on the scale of the Anthropocene, but I want also to emphasise that relationships of kinship and care are equally pertinent within totemic kin groups. The kangaroo people (descendants of the Kangaroo Ancestor), for example, are in a relationship of care with non-human kangaroos (who are the descendants of the same ancestors and thus close kin). The same can be said of the goanna people, and the emu people, the flying-fox people, and so on. Kinship and care are focussed into multispecies family groups as well as into Country.[4]

Every life within Country, and thus within kinship, is surrounded by all this sentience. The implications can be explored through some actual events. One example concerns witnessing the lives of others, and being witnessed. Over a twenty-year period I worked on about eighteen land claims under *Aboriginal Land Rights (NT) Act 1976*, and a few others as well. This *Act* was signed into law on a wave of public support for Aboriginal people and their aspirations. The legislation was specifically intended to benefit Aboriginal people, and at the same time the process was designed to ensure fairness as conceived within Anglo-European law. An Aboriginal Land Commissioner was appointed to conduct an inquiry, traditional owners offered evidence of ownership that was relevant under the terms of the Act and that evidence was tested under cross-examination; persons or institutions (such as the NT government) who perceived that a successful claim might be detrimental to their interests were allowed to make their case for detriment, and the Land Commissioner made his findings. In every claim brought before the Commissioner at least some portion of the land under claim was returned to the traditional owner.

Any legal process that works across cultural differences requires that people be able fully and fairly to articulate their position. In the course of the first few land claim hearings it became clear that the judicial process was actually undermining Aboriginal people's capacity to articulate their position and bring forward their evidence.[5] The Land Commissioners drew on a principle which is firmly embedded in Anglo-Australian legal thought, and to which none of the opponents to the claim could reasonably object: natural justice. Aboriginal people do not in my experience articulate this abstraction, but they are entirely familiar with the concept of a 'fair go,' and they were confident that given a fair go they could explain to an outsider their system of belonging to Country.

In order to ensure that the claimants received a fair hearing, various legal practices had to be modified, and over a period of two decades a new body of practice came into being. The specific practice that concerns me here is that evidence was given on-site. It was presented in Country and in the presence of many of the sacred and historical sites being discussed. The judge sat on the ground with the Aboriginal witnesses, walked through Country with them and listened to stories that were confronting and challenging. The court 'went bush' in every sense of the term.

Part of the reason for on-site evidence was that people's authority to speak is located in Country. To speak while standing or sitting in someone else's Country, as would be the case for most people if all the hearings were held in town, is to speak out of turn (in Aboriginal Law). Claimants were inhibited and uncomfortable. There was a deeper reason, perhaps less accessible to Western law but full of significance for Aboriginal claimants: there is always an 'audience'. Mary Graham's precept, mentioned above, 'you are not alone' says this perfectly in four words. One's authority to speak is authorised by the Dreamings and the ancestors who are in Country, and by Country itself. When one speaks, ancestors and Country are listening. Other living beings may well be listening too. Most assuredly Dreamings are paying attention. It would not be Lawful to speak deeply about matters affecting Country without including all the sentient beings who are themselves living inside of Country.

As a member of the kindred, humans, like all others, have their proper place. The Land Claim process within which Aboriginal claimants ('traditional owners' in legal parlance) insisted on including Country as part of the proceeding show another side to living within one's place. The precept 'you are not alone' has this corollary: your life bears witness to the lives of others, and your life is witnessed by others. You are always engaged, always responsible.

Many of us 'outlaws' would tread the Earth differently if we realised, or would remember, that our actions were being witnessed by others.

Gifts

Ambelin and Blaze Kwaymullina write about creation. In their words, Creation is part of everything that is, and Creation also, of course, has an originary moment: the ancestors, or Dreamings 'who made the world and continue to live within it'. Furthermore, creation is for life: 'the gift of law was to show all life how to sustain Country' (Kwaymullina and Kwaymullina 198).

Law is a gift. This means that life is all gift. There is no entitlement. Creation, like birth, arrives not because 'we' (humans, Western folk, whatever) are so exceptional, so deserving, so much a gift to the universe. The philosopher Jim Hatley is doing wonderful work on creation, and he writes that there is simply no autonomous right to be born, no autonomous right to be brought into life. The embodied life one lives is only ever a gift. The goodness of creation, Hatley says in his complex analysis of Genesis, arises as one is called into relationships, and thus is called to participate in creation (Hatley 253–278). Rather than entitlement there is gift, participation, responsibility, as discussed in the context of kinship. And – the gift is always coming. Every moment is a gift. The food we eat, the water we drink, and every breath of life – it is all gift. Linda is fond of saying that creation is on-going. I am fond of saying that Country is always coming. The same is true of the gifts through which creation is continuous and Country keeps coming. They are always coming forth, always in motion.

So Country is not 'ours' as the government says in its literature on 'Caring for our Country,' as if it were some sort of entitlement or as if we were the boss. Country is an intergenerational, interspecies gift of life. The gift can't be paid off, that would not only be impossible, but is silly even to think of. The ethics of the gift entail that the gift be cared for and handed on to succeeding generations. The gift is respected and cherished. Indeed, there are so many gifts that it is far more appropriate to use the plural – gifts. And to do so helps us pull the concept of gifts back from abstractions and thus to remember that we and everything else only ever live through the always-coming-forth of gifts (see Derrida; Smith). The philosopher Derrida inevitably comes to mind when a person such as myself thinks of 'gift,' and his words are helpful up to a point. He reminds us that the gift is not about exchange, it is only ever an unquantifiable, unpayable debt. In the context of Country and Law, gifts are about responsibility and participation, about celebration and on-going creation.

Once we start thinking 'gift' and thinking 'creation' we experience invigorated horror about the Anthropocene. This current state is not just a bad turn in an otherwise okay history. It is rather the accumulation of millennia of monumentally fatal errors: it involves trashing the gifts of creation. There is destruction of life and livelihood, there are extinctions, there is this life in exile from the law, and participation in a downward cycle of entropy masked as productivity. There is participation in processes the effects of which ramify and continue for generations. The 'slow violence' (Rob Nixon's phrase) of environmental damage and negligence accumulates into great and terrible effects that harm the planet, and harm many, many creatures, and will go on causing harm for generations, perhaps millennia (Nixon).

Mutualism

One of the great Lawmen of the Victoria River region of Australia's Northern Territory was Old Jimmy Mangnayarri. He spoke about Captain Cook and colonisation (as did many people), and one of his questions was why it had all been so hard. Why wasn't mateship offered right from the start? That was what Jimmy Mangnayarri wanted to know: 'Why [Captain Cook] never say: "Oh, come on mate, you and me live together. You and me living together, mates together. You and me can work for Country all the same then"' (quoted in Rose *Dingo Makes Us Human* 194). We had this conversation in about 1992, and I am revisiting his words in light of what we now know about the Anthropocene. Along with the decolonising gesture of reconciliation, I am thinking about how

Old Jimmy was transforming the dyad of coloniser/Indigenous into a triad that includes Country. And how he puts Country at the heart of it all.

This is the absolute issue of our time: how we may work together for Country. This is global, multispecies work. It is the great work that really matters.

When Old Jimmy says that the whole purpose of living together is to work for Country, we hear resonances with the generalisation about keeping things alive in their place. We hear resonances with the Kwaymullinas' words, that 'there is still much work to be done before there can be a true and lasting meeting of minds, hearts and worlds' (Kwaymullina and Kwaymullina 195). In thinking with Linda Hogan, I am particularly attentive to the resonances that emerge, as unintended but welcome harmonies. 'Working together for Country' becomes a 'backbone' for the kind of human action that might be capable of discerning responses to anthropogenic disasters.

Work means actually putting our life energy in the service of Country. This is big. Working *for* Country is fundamentally different to the managerial paradigm expressed in the 'Caring for our Country' programme description discussed earlier. To work *for* Country is to acknowledge the kinship, the generations of nurturance and care, the multispecies energies and flows. There is no entitlement other than this: that every creature lives within webs of responsibility. The Australian government's natural resource management (NRM) site offers a vision of flows of responsibility from government through communities and other managers, toward the protection of the 'natural environment'. It appears to envision Aboriginal people as statistics in a government programme. To work *for* Country flattens the hierarchical structure, includes all participants and the work of their own gifted lives, and puts country at the heart of it. There are webs of care and responsibility that involve everyone. Much endures from generation to generation: the kinship, the responsibilities, the nurturance. In short, Country endures. If sustainability means anything, it surely means endurance from generation to generation. It must mean health Country.

And so, to work for Country is to work for creation, continuity and mutuality. It is to be in the midst of change, and to work with that too. Taking care of Country does not mean engineering it. It means doing our human part while others do their part, and Country's own life-giving capacities have the chance to flourish. Mary Graham's second precept – 'you are not alone' – is central: working together means working with others. Linda writes about the cascades of well-being that ecosystems are capable of when humans don't get in the way; her bison case study is an excellent example. She tells us about how humans can help, and how help can constitute working together when much of the togetherness depends on others doing what their gifted lives enable them to do. It is all backbone.

Together can be understood expansively in another way. I do not think the old man meant that every Indigenous person would have to be working side by side with a non-Indigenous person, and vice versa. Rather, he spoke of sharing – you can be part of my place, and I can be part of your place. Together recognises that our lives are entangled, and that we can be sharing purposes, while working where we are. And, lest we forget, humans are not the only beings working together. In fact, the evidence of the Anthropocene indicates that humans are the main category of creatures who are *not* working together with others. Across much of the world, others are doing the best they can, but humans keep getting in the way.

There are huge issues here, and I will conclude by mentioning just one that is close to my heart: working where we are. About half of all the humans on Earth are now living in urban areas. Not only are humans becoming more urban, so too are many animals. The reasons are many, and go beyond my purposes here.[6] Already there are urban planning efforts to include wildlife, and many urbanites have found ways to develop everyday arts of co-existence. Some such arts simply involve mutual avoidance, but others may, indeed, become convivial as shared places produce a sense of place that is inclusive and mutually beneficial.[7]

My question is: is it possible to re-imagine urban and suburban places as kinds of 'Country'? I absolutely do not want to trivialise Aboriginal Country, and yet I do want to consider the possibility that the Western nature-culture binary presupposes that Country cannot be a city, or part of a city. But is this so? Is it not possible to go beyond the idea of natural areas within urban spaces, an approach which still seems to maintain the nature-culture divide, and to look more thoughtfully at how a city might be reconfigured if the aim of urban life was to inhabit and care for Country. Could the city, and city-dwellers, all of us urban flora and fauna, become part of a story of Country? Could we be working together for Country? Could we find lawful ways of being *where* we are whilst reconfiguring *who* we are so that even in cities we are able to nourish and strengthen our true backbone as kin within the gifts of Earth life?

Notes

1 See www.environment.gov.au/indigenous/index.html
2 See the Australian Government's website at www.nrm.gov.au/
3 The workings of the nature-culture binary have been analysed in numerous publications. I am drawing most particularly on Plumwood and Ingold.
4 In a number of publications I have discussed the implications for self-interest, hunting, and other aspects of 'care' within a system of multispecies kinship, see for example: Rose 'Mates Together'; also Rose 'Death and Grief in a World of Kin'.
5 For a perspective on the Land Claim experience from the point of view of the Aboriginal Land Commissioner, see Gray 'Cabbage Gum Bore'. The volume in which Gray's essay appears contains numerous accounts by Aboriginal men and women.
6 Their habitat is removed, or the suburb is built in their habitat, or cities provide a range of new opportunities that facilitate their lives ... the list goes on.
7 'Urban wildlife' as a term seems both to undermine the nature-culture binary and to sustain it. There is a growing literature on this; see for example Lunney and Burgin; van Dooren and Rose; and Wolch.

Works cited

Derrida, Jacques. *Given Time: I Counterfeit Money*. Trans. Peggy Kamuf. Chicago: University of Chicago Press, 1994. Print.

Graham, Mary. 'Some Thoughts on the Philosophical Underpinnings of Aboriginal Worldviews', *Australian Humanities Review* 45 (2008): 105–118. Print.
Gray, Peter. 'Cabbage Gum Bore', *Take Power like this Old Man Here*. Ed. Alexis Wright. Alice Springs: IAD Press, 1998. 45–47. Print.
Hatley, James. 'The Original Goodness of Creation: Monotheism in Another's Voice.' *Facing Nature: Levinas and Environmental Thought*. Eds William Edelglass, James Hatley and Christian Diehm. Pittsburgh: Duquesne University Press, 2012. 253–278. Print.
Ingold, Tim. *The Perception of the Environment: Essays in Livelihood, Dwelling and Skill*. Abingdon and New York: Routledge, 2000. Print.
Kwaymullina, Ambelin and Blaze Kwaymullina. 'Learning to Read the Signs: Law in an Indigenous Reality', *Journal of Australian Studies* 34.2 (2010): 195–208. Print.
Lunney, Daniel and Shelley Burgin, eds. *Urban Wildlife: More Than Meets the Eye*. Mosman: Royal Zoological Society of New South Wales, 2004. Print.
Muecke, Stephen. 'The Sacred in History', *Humanities Research* 1 (1999): 27–37. Print.
Nixon, Rob. *Slow Violence and the Environmentalism of the Poor*. Cambridge, MA: Harvard University Press, Cambridge, 2011. Print.
Plumwood, Val. *Feminism and the Mastery of Nature*. London: Routledge, 1993. Print.
Rose, Deborah Bird. *Country of the Heart: An Indigenous Australian Homeland*. With Sharon D'Amico, Nancy Daiyi, Kathy Deveraux, Margy Daiyi, Linda Ford and April Bright. Canberra: Aboriginal Studies Press, 2011. Print.
——. 'Death and Grief in a World of Kin', *The Handbook of Contemporary Animism*. Ed. Graham Harvey. Durham (UK): Acumen, 2013. 137–147. Print.
——. *Dingo Makes Us Human; Life and Land in an Australian Aboriginal Culture*. Cambridge: Cambridge University Press (third printing), 2009. Print.
——. 'Mates Together: Dancing with Difference.' *As Others see Us: The Values Debate in Australia*. Eds Vin D'Cruz, Bernie Neville, Devika Goonewardene and Phillip Darby. Melbourne: Australian Scholarly Publishing, 2008. 69–78. Print.
——. *Nourishing Terrains; Australian Aboriginal Views of Landscape and Wilderness*, Canberra: Australian Heritage Commission, 1996. Print.
Smith, Mick. 'Hermeneutics and the Culture of Birds: The Environmental Allegory of "Easter Island"', *Ethics, Place and Environment: A Journal of Philosophy and Geography* 8.1(2005): 21–38. Print.
van Dooren, Thom and Deborah Rose. 'Storied-places in a Multispecies City', *Humanimalia* 3.2 (Spring 2012): 1–27. Print.
Wolch, Jennifer. 'Anima Urbis', *Progress in Human Geography* 26.6 (2002): 721–742. Print.

4 Introduction to Parts II and III

Backbone and Country

Michael Davis

Seeing, moving and walking together in the Anthropocene

The idea of an observatory is not new. As a process, a way of seeing, of imagining and reflecting on our place in the cosmos, the observatory has ancient origins. The notion of an observatory as a place to see things in their entirety, or from a wide purview of inquiry, informs the development of the Humanities for the Environment (HfE) Project and the global HfE Observatories. This idea of observing through the HfE Observatories is the theme for this book *Humanities for the Environment (HfE): Integrating Knowledge, Forging New Constellations of Practice*, acknowledging that, as Joni Adamson points out in her Introduction, 'observation' is not a neutral or value free concept. Using this theme of observing, the chapters in this volume discuss cross-cultural encounters and environmental pasts, sustainable livelihoods and Indigenous futures. They also raise questions about the concept of the Anthropocene, take up discussions around ecocriticism, tangled reef ecologies, narrative ethics and bushfires, and consider the relatively unexamined notion of curating the Anthropocene. The chapters range across geographies, ecosystems, climates and weather regimes; from icy Arctic landscapes, to the tropics of north Queensland and the Great Barrier Reef, and to urban 'laboratories' on a university campus in Phoenix, Arizona. In these diverse settings and contexts, this volume showcases how, using multidisciplinary approaches to 'integrating knowledges,' educators, photographers, scientists and artists across the humanities, social and hard sciences can help foster awareness about the crisis faced by anthropogenic processes on the planet.

In my own chapter in the book, linking observation and seeing with walking, I employ the phrase 'walking together' to suggest a process of collaborative observing, seeing and reflecting that might be seen as one unifying theme of the collection. Rebecca Solnit has also explored this idea, writing in *Wanderlust* that when we are walking, we are in 'a state in which the mind, the body, and the world are aligned' (Solnit 5; Davis, this volume; see also Davis 'Entangled Tropical Knowledges'). In this reflective mode, as our feet move along the ground, we are connecting with the land, or with 'Country,' to use the formulation in Deborah Rose's chapter 'Country and the Gift.' In walking, our bodies are supported by our backbone, which enables structure and balance to our locomotion. In Linda Hogan's chapter 'Backbone: Holding Up Our Future,' balance is also found through the Chicaza (Chickasaw, North America) words 'tish' and 'punyu,' words that together form part of these peoples' ancient knowledge systems. Backbone and Country, with their ideas of balance, movement and knowledge, connect to 'walking together,' which forms a metaphor for 'new constellations of practice,' the central theme for this volume. In 'walking together,' peoples, things, multiple species and disciplinary practices become entangled, and this enables new ways of thinking in the Anthropocene that unsettle received hierarchies, uniformities and hegemonies.

Like Solnit, seeking and finding connections and movements between things also underpinned the work of Greg Dening, Australian historian and anthropologist of the Pacific. Dening had a fascination with beaches and islands. But to him these were not 'fixed' places; he was more interested in the flows, journeys, crossings and stories that connected beaches and islands, and in the peoples living in, and moving between them. In *Beach Crossings: Voyaging Across Times, Cultures and Self* he discussed the flight of a migratory bird, the short-tailed shearwater, as a metaphor for what he described as the 'encompassing of Oceania'; the 'enveloping of space with knowledge,' as a way of 'imprinting the island-seascape with spirit and life – human, animal, vegetable' (16).

The Tongan writer Epeli Hau'ofa was also interested in connections, and he saw the Pacific not in terms of individual, isolated islands, but as a vast region – Oceania – that was the sum of its places, peoples and the movements between and among the islands. In the context of Dening's and Hau'ofa's notions of flows and relationships, the chapters in this book are concerned with practices and disciplines that connect oceans, continents, islands and peoples, and which find articulation through seeing, moving and 'walking together.' But these connections have not always been readily made, and they have, over time, become fragmented and compartmentalised into disciplines and specialisations. This separation permeates many areas of human thought and endeavour, including in the ways that Europeans have continued to depict and represent Indigenous peoples' cultures, societies and knowledge systems (see Davis *Writing Heritage*). 'Walking together' offers a focus for thinking about the theme of this book as 'new constellations of practice,' as ways of re-connecting these otherwise fragmented and disparate concerns, knowledges and domains. But as the chapters show, collaboration, observation and seeing together as 'movement' also embraces other, seaborne forms, such as canoeing and paddling.

Part II: Backbone

The notions of balance, movement and knowledge embedded in the use of the term 'backbone' articulated in Linda Hogan's chapter of this volume offer a fulcrum for thinking about the various 'new constellations of practice' discussed in the chapters in this section of the book. The first chapter in this part deals with islands – those in the Pacific and the Caribbean. Karen N. Salt, in 'Twilight

Islands and Environmental Crises: Re-writing a History of the Caribbean and Pacific Regions through the Islands Existing in their Shadows' (Chapter 5) investigates the histories of what she calls 'twilight islands,' small, mostly non-human inhabited spaces and biodiversity hotspots that represent forgotten lands with deep histories and tragic tales of human warnings about dispossession (between and amongst multiple species). Salt argues for an interdisciplinary focus that can comprehend the ways that all islands are deeply imbricated in the history of human/nature relations. Instead of being tangent to environmental and ecological crises, twilight islands sit directly in the midst of them; yet they continue to be shaded in 'twilight' because they exist within a set of historic spaces that reveal the shifting terrain of ecological protection. Continuing a theme around islands and water, in Chapter 6, 'Seaweed, Soul-ar Panels and Other Entanglements,' Giovanna Di Chiro writes about the influences, collaborations and 'multi-sited natureculture entanglements' that have been crucial in shaping her work in environmental activism and writing over several decades. Reflecting on her experience in marine science on Hawai'ian coral reefs, and her involvement in a 'seaweed sisterhood,' Di Chiro describes her entanglement in reef ecologies, showing how these enabled her to challenge the hegemony of conventional marine science, and see it instead as a complex intertwining of knowledge, science and culture.

This notion of a more coherent relationship between nature and the human spirit courses throughout all the chapters in this collection, but is particularly noticeable in the chapters by Kyle Whyte and Joni Adamson. More coherent forms of relationships between humans and nature were found well before the twentieth century, in both Indigenous and Western cultures. For example, seeking a unifying vision of nature, the German explorer and scientist Alexander von Humboldt wrote in his 1844 Preface to the first volume of his book *Cosmos*, which was published in 1845:

> The principal impulse by which I was directed was the earnest endeavor to comprehend the phenomena of physical objects in their general connection, and to represent nature as one great whole, moved and animated by internal forces.
>
> (7)

Humboldt's reference here that in seeking to understand nature he was 'moved and animated by internal forces' was likely indicating his interest in seeking a unity between nature and imagination. His vision of nature was as much a poetic one as it was based in physical presence, and held some sway until around the late nineteenth century (see Adamson 'Source of Life' 253–255; also Wulf). Until then, thinkers, explorers, and all those who sought to inquire into, and accumulate information about the world and its peoples, still tended to work within a more unified domain of knowledge than was the case in later decades. Humboldt's sense of curiosity and inquiry is central to the vision of the environmental humanities, as Joni Adamson has argued ('Humanities' 137).

Many of the chapters in this collection also show how thinking within and beyond the Anthropocene must attend to notions of vulnerability, fragility and also resilience. In this context the voices of Indigenous peoples are vitally important, and their distinct cultures and histories, and deeply embedded knowledges, figure prominently throughout this collection. In Chapter 7, 'Is it Colonial Déjà Vu? Indigenous Peoples and Climate Injustice,' Kyle Whyte focuses on his own people the Neshnabé, or Citizen Potawatomi Nation of North America, who lived as an ecologically mobile, culturally distinct and politically independent society that harvested plants and animals from complex ecosystems and who stewarded their habitats and cultivated crops across an area of thirty million acres in the Great Lakes region of North America. In policy and academic literatures, Whyte observes, current theories of vulnerability and climate risk often claim that climate injustice against global south populations, which include millions of Indigenous peoples, result primarily from 'bad luck'. He links colonialism, climate justice and the Anthropocene to show that climate injustice is not an unprecedented future for Indigenous peoples; rather it is a continuation of patterns of infliction of anthropogenic environmental change that have been part of settler colonialism for quite some time.

Another group of Indigenous peoples of North America, the O'odham, are discussed in Joni Adamson's chapter (8), 'Gathering the Desert in an Urban Lab: Designing the Citizen Humanities'. She introduces her students to the sophisticated astronomical and ethnobotanical knowledges of the O'odham and asks them to engage in 'new constellations of practice in the humanities,' including storytelling, observation, drawing, images, words, time-lapsed film and photography, which might help others to 'see' a nonfamiliar species. Adamson turns her own university's campus into an 'urban lab' inspired by both ancient observatories and by modern international research platforms such as Future Earth, then invites her students to join her in piloting new forms of 'citizen humanities' that would enrich what we mean by 'citizen science' and bring both to the creation of livable futures. She discusses innovative pedagogical templates that are made easily available on digital platforms for other teachers and communities in order to 'scale up' our response to climate change and extinction.

In a different form of knowledge production, Mark Klett and Tyrone Martinsson, in Chapter 9, 'Environmental Rephotography: Visually Mapping Time, Change and Experience,' discuss their work using 'rephotography' or repeat photography, to document and map changes in environments and climates. The technique of rephotography is among the most important photographic techniques and 'constellations of practice' for addressing deep time, environmental change and human perception. In this process, one or more photographs made at different times are repeated to form points of comparison, anchoring historical documents with physical locations. Rephotography forms a dialogue with time, history and memory, and can contribute to our understanding of the environment and its past, affecting decisions regarding the paths we choose to take into the future. Klett and Martinsson identify physical and cultural changes in the Western United States and in Svalbard, Norway, in the Arctic, making a record of desert drought, receding glaciers and other effects of climate change. By merging creative practice, research and scientific interpretation, and photography, they provide perspectives on both deep geological time and rapid present-day change.

The theme of balance and harmony in Linda Hogan's introductory chapter 'Backbone' is also one that underpins Michael Zimmerman's work in Chapter 10, 'Integral Ecology in the Pope's Environmental Encyclical, Implications for Environmental Humanities'. Zimmerman's chapter, as with Hogan's, can also be read as a call for a shift in action on a larger scale. In Zimmerman's argument, religious leaders like the Pope, by becoming thought leaders for thousands, will be needed to help us 'scale up' for effective response to climate change. Zimmerman critically examines the philosophy of 'integral ecology,' or the version of it that takes up one chapter of Pope Francis' 2015 encyclical *On Care for Our Common Home*. He argues that an 'integral ecology' sees insights from the natural sciences and engineering as critical to understanding environmental issues. Integral ecology also takes the position that relying solely on technical assessments for policy decisions amounts to technocracy, or the authoritarian rule of experts. In this chapter Zimmerman takes the measure of Francis' proposal for an alternative to planet-destroying capitalism which emphasises local, Indigenous values, community-oriented work and products, and appropriate care for humans and non-humans.

Part III: Country

The deeply embedded and multi-faceted connections to land and place that Indigenous peoples have, encapsulated by Deborah Bird Rose as 'Country,' forms a central theme for Part III of this collection. This notion of 'Country' allows us to move beyond linear concepts of time, place and connection, to embrace a different way of thinking about connection, relating and belonging. If 'Backbone' can be a metaphor for balance and harmony in movement, then 'Country' enables thinking about movement through place and time, and also how we 'sit on,' or walk through place as 'Country'. This brings us back to the theme of 'walking together,' a phrase that can be summoned throughout this volume, and certainly in the chapters that comprise Part III. Walking together on country – country connoted as much as about land as about sea – enables a thinking in terms of connections, and of 'wholes' rather than 'isolates'. It invites observation: seeing and moving together in collaboration.

'Walking together' as a metaphor embracing all kinds of movement, including seaborne ways such as paddling, canoeing, sailing and swimming, can also be used as a way of envisaging, or observing things in a coherent, unifying way. This returns us to Epeli Hau'ofa's vision, which was also of 'wholes,' a notion which he encapsulated in his writing about Oceania. Reflecting the perspectives of the peoples of this region, he stated:

> their universe comprised not only land surfaces, but the surrounding ocean as far as they could traverse and exploit it, the underworld with its

> fire-controlling and earth-shaking denizens, and the heavens above with their hierarchies of powerful gods and named stars and constellations that people could count on to guide their ways across the seas.
>
> (152)

This evokes the way that Pacific and Oceanic peoples view their worlds, as interconnected domains encompassing the physical realm, as well as that of the gods, spirits and ancestors. Hau'ofa's Oceania, as a 'sea of islands,' denotes a 'more holistic perspective in which things are seen in the totality of their relationships' (152–153). In this view, islands, lands, resources and environments cannot be conceived of separately from the people who inhabit the region. Hau'ofa was writing this in 1994. In today's Oceanic world, connectivity between and among islands and places is transformed by the considerable advances in digital communications. The Humanities for the Environment (HfE) Australia–Pacific Observatory, a node of the interlinked HfE Observatories that encompass those in North America, Asia, Europe and Africa, contributes to this interconnectedness that is at once digital, and geographic.

In the spirit of Hau'ofa's writing, and also Dening's, two of the chapters in this section, first and last, (Huang and Rapongan, Chapter 11 and DeLoughrey, Chapter 16) turn their gaze towards watery matters; and these chapters 'bookend' this part, drawing together reefs, oceans and islands, and connecting places and peoples. Hsinya Huang and Syaman Rapongan, in Chapter 11 'Radiation Ecologies, Resistance and Survivance on Pacific Islands,' discuss the forms of resistance and collaboration that brought together environmental humanists, writers, artists and local groups, to resist the violence of Pacific nuclear histories. Their chapter illustrates how Indigenous and local voices were able to organise a struggle and a counter movement to challenge nuclear colonialism, and how those actions have also been taken up by Pacific Indigenous writers. Co-author Syaman Rapongan, a Tau elder, has written elsewhere about how the Tau continue their age old skills in fishing, canoeing and boat-building. Paddling from island to island throughout Austronesia to organise a movement to resist nuclear militarism, Rapongan is, in this co-authored chapter, also concerned with forms of movement, in this case seaborne movement.

Coral reefs and entangled reef knowledges feature in several of the chapters in Section III, including those by Michael Davis and Elizabeth DeLoughrey. The Australian Great Barrier Reef figures in my chapter (12), 'Walking Together into Knowledge: Aboriginal/European Collaborative Environmental Encounters in Australia's North-East, 1847–1849.' Although not explicitly present in my discussion, this Reef forms a 'backdrop' for my analysis of encounters that took place in the mainland rainforest abutting the Reef, between local Aboriginal people and British explorers and naturalists in nineteenth century Australia. These encounters, I argue, can be read as moments of cooperation and collaboration, possibly also friendship, during which people from different cultures met in 'conversations' about plants and animals. The chapter deploys the phrase 'walking together' as a framing device for an analysis of these encounters, which, it is suggested, encapsulate instances of 'multispecies' engagement. In this chapter I juxtapose those encounters with another 'walking together' event, that took place in Paris in December 2015 when nearly 190 nations met to agree to an international approach to mitigate the effects of global climate change (see also Joni Adamson's Introduction, this volume). The outcome of the Paris meeting was the result of many years' thought and action, including the kind of 'integral ecology' espoused by Pope Francis in his encyclical, and discussed by Michael Zimmerman in Chapter 10.

The process of walking or 'walking together' is one that is especially apt for thinking about the Anthropocene, human induced climate change, and for envisaging collaborations and connections, and entanglements between humans, other species and the world of objects. It also offers a useful frame for, or is guided by the philosophy of the HfE Observatories, stimulating as these seek to do, a far greater engagement between the disciplines, particularly between the human and social sciences, and the so-called 'hard' sciences.[1]

Vulnerability, resistance and resilience are recurring themes in the Anthropocene, and are central to several chapters. As in Kyle Whyte's chapter, Melbourne based Aboriginal scholar and award-winning novelist Tony Birch is also attentive to the histories and struggles of Indigenous peoples, and to the vital role of Indigenous peoples in Anthropocene discussions and discourses. In Chapter 13, 'The Lifting of the Sky': Outside the Anthropocene,' Birch argues that Indigenous peoples' knowledges, cultures and histories have tended to be sidelined in Anthropocene discourses. He discusses the contested nature of the Anthropocene in his provocative critique, and argues that despite the vast amount of scholarship published on the Anthropocene, there is still relatively scant attention given to the role of Indigenous peoples and their knowledges, despite the fact that Indigenous people produce the smallest carbon footprints on the planet, and yet many communities are most vulnerable to experiencing the immediate and devastating impacts of climate change. Birch argues that new alliances and cooperative conversations are required to deal with climate change. Indigenous people must play a key role in these shifts, both for the benefit of local communities and national and global interests.

Vulnerability, fragility, and the complex ways of being in Country are powerful themes in Kate Rigby's discussion of bushfires in Australia. In Chapter 14, 'Literature, Ethics and Bushfire in Australia,' Rigby looks at entangled interdisciplinary 'conversations' in an eco-critical perspective on the devastating fire storms that regularly invade south-eastern Australia. Her discussion is contextualised within a longer history of transdisciplinary conversations in the environmental humanities in Australia since the late 1990s. Rigby's chapter exemplifies the ways in which ecocriticism has contributed centrally to the environmental humanities, and she has had a key role in Australia's contributions to both ecocriticism and the environmental humanities.

Ideas about Country also figure in Kirsten Wehner's chapter (15), 'Placing the Nation: Curating Landmarks at the National Museum of Australia,' in which she discusses museums and museology, and curatorial practices and discourses in an

Anthropocene inspired framework. In her critically engaged discussion, she considers how a National Museum of Australia exhibition called *Landmarks: People and Places Across Australia* expresses curatorial efforts to re-shape history museum practices in the context of the Anthropocene, in particular through disrupting modernist tropes of the nation-state. Wehner's chapter engages with Indigenous knowledges and ideas about 'Country,' as she considers how the *Landmarks* exhibition's curators' development of a place-based history of Australia, responded to Aboriginal and Torres Strait Islander peoples' concepts of 'Country,' as well as scholarship in the environmental and ecological humanities and material culture studies. Curation as discussed by Wehner is a good example of a 'new constellation of practice' and is the subject of a forthcoming volume in Routledge's Earthscan series, entitled *Curating the Future: Museums, Communities, and Climate Change.*

Returning to the watery theme and Hau'ofa's connected islands, in Chapter 16, 'The Oceanic Turn: Submarine Futures of the Anthropocene,' Elizabeth DeLoughrey maps out an 'oceanic turn' in the humanities that she has helped to catalyse. She asks what an orientation towards 'sea ontologies' implies for a re-envisaging of temporality and aesthetics in the Anthropocene era. Focusing on island and ocean-centred regions, the Caribbean, and briefly, Australia's Great Barrier Reef, DeLoughrey's analysis of interdisciplinary engagements in critical ocean studies allows a reading that disrupts ideas about stability, linearity and fixity, towards transformations in our thinking about time, space and knowledge. The Great Barrier Reef, which DeLoughrey identifies as 'one of the indicators' for the crisis faced by coral reefs in the Anthropocene, is an example of a 'multispecies assemblage' (see also Ogden et al.; Davis, Chapter 12 this volume), a notion that allows us to envisage this vast area of rich biodiversity as a rich, entanglement of the human and the natural. Here DeLoughrey refers to Iain McCalman's recent work *The Reef: A Passionate History*, which itself considers the complex histories of engagement between the Reef's human stories and its science. McCalman, together with Libby Robin and Deborah Bird Rose, is among the people who have been at the forefront in establishing the Humanities for the Environment in Australia and connecting it globally, a process that has included Iain's leadership in establishing the Sydney node of the HfE Observatories. In a re-theorising of oceanic studies, DeLoughrey's chapter illustrates the kind of critical multidisciplinary thinking that articulates themes of 'walking and moving together,' disrupting as it does, received ideas about place, time, and ways of seeing.

The chapters in *Humanities for the Environment (HfE): Integrating Knowledges, Forging New Constellations of Practice*, in their diverse ways, all stimulate ideas in which multiple things, actors, objects, places and knowledges, practices and disciplines are working together, thus articulating powerfully an Anthropocene focused call for 'integrating knowledges'. The meeting of the COP in Paris in December 2015 exemplified a 'walking together,' taking this as a metaphor for many different ways of moving together, that seeks to bridge disciplines, geographies and ideologies, bringing together a wide range of participants into conversation for the Earth. *Humanities for the Environment* (HfE) seeks to extend the ways we think about human culture, reframe understanding of life, social and cultural practices, values and resources as they present the networked research of the North American, the Asian–Pacific and Australian–Pacific Observatories.

Note

1 The H*f*E Common Threads were written one year into the grant period by H*f*E Conveners, Principle Investigators and Key Researchers who met at the 2014 international meeting of CHCI in Hong Kong. See http://hfe-observatories.org/common-threads/

Works cited

Adamson, Joni. 'Humanities.' *Keywords for Environmental Studies*. Eds Joni Adamson, William A. Gleason, and David N. Pellow. New York: New York University Press, 2015. 135–139. Print.

——'Source of Life: Avatar, Amazonia, and an Ecology of Selves'. *Material Ecocriticism*. Eds Serenella Iovino and Serpil Opperman. Bloomington, Indiana: University of Indiana Press, 2014. 253–268. Print.

Davis, Michael. 'Entangled Tropical Knowledges: Towards a Poetics of Knowledge and Place-Making in Nineteenth Century Voyaging Narratives'. Proceedings of the Tropic of the Imaginations Conference 2015. Special issue of *etropic* 15.1 (2016). Web.

——*Writing Heritage: The Depiction of Indigenous Heritage in European-Australian Writings*. Melbourne: Australian Scholarly Publishing, 2007. Print.

Dening, Greg. *Beach Crossings: Voyaging Across Times, Cultures and Self*. Carlton, Victoria: The Miegunyah Press, 2004. Print.

Hau'ofa, Epeli. 'Our Sea of Islands', *The Contemporary Pacific*, 6.1 (Spring 1994): 147–161. Print.

McCalman, Iain. *The Reef: A Passionate History*. Melbourne: Viking/Penguin Books, 2013. Print.

Newell, Jennifer, Libby Robin and Kirsten Wehner (forthcoming). *Curating the Future: Museums, Communities, and Climate Change*. Routledge. Print.

Ogden, Laura, Nik Heynen, Ulrich Oslender, Paige West, Karim-Aly Kassam and Paul Robbins, 'Global Assemblages, Resilience, and Earth Stewardship in the Anthropocene', *Frontiers of Ecological Environment* 11.7 (2013): 341–347. Print.

Rapongan, Syaman. 大海浮夢 [*The Drifting Dream on the Ocean*]. Taipei: Union Publisher, 2014.

Solnit, Rebecca. *Wanderlust: A History of Walking*. New York: Viking Penguin, 2000. Print.

von Humboldt, Alexander. *Cosmos: A Sketch of the Physical Description of the Universe*, Vol 1. Trans. E. C. Otté. Baltimore and London: The Johns Hopkins University Press, 1997 [1858 English edition]. Print.

Wulf, Andrea. *The Invention of Nature: The Adventures of Alexander von Humboldt, The Lost Hero of Science*. London: John Murray, 2015. Print.

Part II
Backbone

5 Twilight islands and environmental crises

Re-writing a history of the Caribbean and Pacific regions through the islands existing in their shadows

Karen N. Salt

> The dusk was a raucous chaos of curses, gossip,
> And laughter; everything reformed in public,
> But the voice of the inner language was reflexive and mannered,
> As far above its subjects as that sun which would never set
> Until its twilight became a metaphor for the withdrawal of empire
> And the beginning of our doubt.
>
> Derek Walcott[1]

Introduction

In 1970, St Lucian born poet and playwright Derek Walcott published a moving 'overture' to his play, 'Dream on Monkey Mountain,' that he dubbed 'What the Twilight Says'. In this lyrical essay, Walcott traverses through the landscapes of colonisation, transliterating the lure of imperialist languages formed from the dispossession of islands and their seas. Haunting and emotive, this short piece chronicles the dangers of living within twilight. According to Walcott, twilight represents the transitional potential of the time after the dark of empire and colonialism. Rather than see twilight as a hopeful condition, Walcott envisions it as a violent process of control that appropriates the language of self-determination for its own machinations.

More than twenty years after the publication of 'What the Twilight Says,' Walcott returned to the notion of twilight in the Antilles in his Nobel lecture, 'The Antilles: Fragments of Epic Memory,' but frames it not as a process, but as an occluded state of seeing (or not seeing) that has political and environmental implications. He challenges critics and other interested observers to remember that the Caribbean nourishes and sustains its people through its very roots. As organic beings sustained by their environment, Caribbean people face numerous threats of destruction as 'every day on some island, rootless trees in suits are signing favourable tax breaks with entrepreneurs, poisoning the sea almond and the spruce laurel of the mountains to their roots'. Prophetically, Walcott anticipates a coming morning in which governments might wake up and wonder, 'what happened not merely to the forests and the bays but to a whole people' (Walcott *The Antilles* n. pag.).

Walcott's musings were my introduction to the concept of twilight – as both a process of continued control and as an occluded state of seeing (or not seeing). Navassa Island, a two square mile U.S. controlled territory 35 miles southwest of the coast of Haiti, was my first physical and material example of a 'twilight' island inhabited by no people. A site of occasional subsistence fishing along its coastal waters, Navassa is a small mostly coral and limestone island with cliffs, a number of endogenous species, and no fresh water. Claimed by the U.S. in the 1850s (as an island where guano mining wreaked havoc on the island's ecology and the lives of the mostly conscript African American enforced labourers), Navassa has also lived its life as a U.S. military site before becoming, since the late 1990s, a protected site that forms part of the U.S. Caribbean Wildlife Refuge. I have written about this history in other places, so I will not re-tread that ground (Salt 267–286). What is relevant here is Navassa's geo-political links to other uninhabited islands.

Although I encountered Navassa through my work on black sovereignty, Haiti, and the nineteenth-century Atlantic world, I have since collected evidence of countless twilight islands in the Caribbean and around the world. I define twilight islands as small, mostly nonhuman inhabited spaces and biodiversity hotspots that represent forgotten lands with deep histories and tragic tales of human warnings about dispossession (between and amongst multiple species). As is the case with Navassa, many of the global interactions with these islands have been framed through the lens of racial difference that has cast the local or regional inhabitants as less able, less political or even less human as compared to people or empires in the North Atlantic or the more amorphous 'west'.

Because no people live there, historians typically do not study twilight islands, other than as outposts of imperial histories that recount stories of dispossession within larger island systems.[2] In leaving these smaller, nonhuman inhabited islands outside of the larger arc of coloniality and empire within histories of these lands and seas, crucial formal and informal networks of control have been left out of narratives and studies of island territories, as well as their presence across the globe.

As significant nodes in the resource consumption and global distribution of certain materials, twilight islands have critically sustained and adapted these processes. However, historians do not consider them as a set of sites that participate in a range of complex global processes. Instead, each island is typically only globally linked through the mechanisms of extraction, distribution and consumption of particular resources, such as guano. As a consequence, researchers have typically not focused on the ways these locales have been and continue to be similarly consumed, exploited and even protected by a range of powerful national and international non-governmental players and business conglomerates. Focusing on them as a larger and more global set – using interdisciplinary methods and humanistic inquiry – allows us, as members of the global public, to make better decisions about their resource use and the trade laws that encompass them.

For many of these islands, dwelling in twilight means living with slow (and unnoticed) ecological destruction.

The temporality of twilight islands and their environmental processes recalls Rob Nixon's theoretical riffing on 'slow violence'. In his important and timely text, *Slow Violence and the Environmentalism of the Poor*, Nixon, one of the most influential voices in the environmental humanities, argues that 'attritional catastrophes that overspill clear boundaries in time and space are marked above all by displacements – temporal, geographical, rhetorical, and technological displacements that simplify violence and underestimate, in advance and in retrospect, the human and environmental cost' (7). The human and environmental costs of these displacements? History. Nixon continues: 'Such displacements smooth the way for amnesia, as places are rendered irretrievable to those who once inhabited them, places that ordinarily pass unmourned in the corporate media' (7). He challenges critics and activists, alike, to 'adjust our rapidly eroding attention spans to the slow erosions of environmental justice' – a call that can be readily applied to twilight islands (8).

This chapter responds to Walcott's musings on twilight and Nixon's challenge to consider the 'unmourned' places that surround us. It does this by detailing the contours and sovereign conditions of twilight islands (and their hauntings) before offering two examples of the processes that have been 'tested' on/within them. As a whole, this chapter argues for more interdisciplinary focuses on island systems in order to comprehend the ways that islands with no human populations are deeply imbricated in the history of human + nature relations. The global import of these spaces became apparent when I joined Hsinya Huang and other environmental humanists, artists and scientists, including Joni Adamson, Elizabeth DeLoughrey and Linda Hogan, at a conference in Taiwan. Later, we organised the Islands and Oceans Ecology Research Network, funded by National Sun Yat-Sen University, in Kaohsiung, Taiwan. This network draws together researchers interested in the trans-oceanic cultures and ecologies of Austronesia and Taiwan, the Caribbean, Scotland and the Pacific. During our official launch of the network, literary critic Huang and geographer Huei-Min Tsai planned a 'knowledge sharing' excursion to the 2013 International Geographers Union's Commission on Small Islands conference in the Penghu Archipelago, Taiwan, where we met with other ecologists, scientists and policymakers who were part of the Network.

Discussions during this gathering greatly informed my theories and ideas about twilight islands and these activities have directly contributed to the formation of the Asia–Pacific Observatory which is part of the international Humanities for the Environment Project.[3] Critically, these networked conversations are situating the humanities within global conversations on environmental futures – a conversation that I participate in through my research on twilight islands.

My work, and the wider work within the environmental humanities, challenges the occlusion of such spaces by highlighting their deep histories.

Rather than argue for the importance of all island ecosystems, this chapter threads nonhuman inhabited systems together in order to reveal not just environmental catastrophes on a large scale, but crises that I deem eco-governmental and racio-ecological at the small scale. These crises, although different in form, motivation and scale, should be analysed together to form a political nexus of global exchange that resists the intentional obfuscation of their important role in global systems. What this work reveals, as does the environmental humanities as a whole, is the ways that these islands exist in the shadows.

I have discovered in my research that many twilight islands are surrounded by a troubling constellation of environmental, racial, economic and political actors who interact with these spaces in conflicting and competing ways. At times capitalistic, and at other times protective, these interactions include forms of control that offer up a 'soup of signs,' borrowing from Cuban novelist and critic Antonio Benítez-Rojo's *The Repeating Island*, that demand multidisciplinary examinations in order to lift the veil of occlusion and fully see them, as well as map their effects (2).

As environmental humanists working on the interplay between politics and ecologies have been arguing for over a decade (See Adamson, "Introduction", this volume), making these effects legible remains a challenge. Writing in *Postcolonial Ecologies: Literatures of the Environment*, Elizabeth DeLoughrey and George B. Handley suggest that challenges to legibility should include more than just imperial and colonial concerns. They argue that although a number of environments have been ravaged by capital, fantasies of racial superiority and pursuits of plunder, they have also experienced the damning effects of

> modernization schemes enforced by the IMF and World Bank such as the construction of hydroelectric dams, the use of agricultural chemical fertilizers and patented seeds, as well as other resource extraction initiatives like deforestation, mining, and the liberalization of internal markets.
>
> (16)

The processes that DeLoughrey and Handley link to postcolonial ecologies also impact twilight islands.

Critical geographers, such as Sian Sullivan, have given significant attention to nonhuman sites. Sullivan's recent essay in Antipode urges a recognition of the 'revolutionary shift in discourses and practices regarding [...] nonhuman natures' (212). For Sullivan, this shift has engendered new relations amongst nation-states, banks, international non-governmental organisations, environmentalists and conservationists who have built on 'extant understandings of [the environment] as property,' and extended this articulation to now include 'radical ways to release new nature "values" that can be traded, invested in and speculated on via conceptual and capitalised conversion into the commodity form' (212). Sullivan's critique of 'banking nature' expands Nixon's theories of slow violence, but tends to read nonhuman nature as a thing now channeled into commodity flows as opposed to considering it as a place impacted by those flows of capital – as I argue is the case for twilight islands. Combining the theoretical ideas of Sullivan, Nixon, Walcott and other environmentally engaged thinkers into a more critical interdisciplinary environmental humanities enables the generation, as Stephen Hartman asserts, of 'new knowledge' that might confront the 'expansive and knotty' ecological challenges (and hauntings) that twilight islands currently face (Hartman).

Islands bathed in twilight: a haunting consideration

Avery Gordon, in *Ghostly Matters*, notes that haunting is an apt way to describe 'abusive systems of power [that] make themselves known and their impacts felt in everyday life, especially when they are supposedly over and done with (slavery, for instance) or when their oppressive nature is denied (as in free labor or national security)' (xvi). Gordon pushes the idea of haunting, time and states of being further by suggesting that hauntings also cause tangible damage to actual persons – a point that I would argue also pertains to entire ecosystems. She notes:

> Haunting raises specters, and it alters the experience of being in time [...] The whole essence, if you can use that word, of a ghost is that it has a real presence and demands its due, your attention. Haunting and the appearance of specters or ghosts is one way [...] we are notified that what's been concealed is very much alive and present, interfering precisely with those always incomplete forms of containment and repression ceaselessly directed toward us.
>
> (xvi)

Gordon's notion of haunts casts the world in new light, but tends to consider the 'matter' of ghostly matters as something that is not grounded in the physical. DeLoughrey places haunting back into the environment, and into materiality, or matter, arguing in a recent essay that seas and oceans contain significant haunts. Within these waters are 'place[s] where the haunting of the past overtakes the present subject' (DeLoughrey 703). For me, the key in understanding these hauntings and their ties to twilight islands lies in how we chart their existence. And they exist in droves. Dotted across the Caribbean Sea (and in other seas and ocean waters on the planet) lie a host of islands – isolated, unremarkable spaces that play significant roles in political and economic battles between territories. Although some of the battles are new, many are old – and haunted.

Two competing ideas come from the word haunt: a noun that demarcates a place that one frequents and a verb form that suggests the appearance of a ghost or supernatural entity (the ghostly matter at play in Gordon's work). In many ways, both forms are apt descriptors of the complex dance of de-familiarisation, re-familiarisation and consumption tied up in the relationships between the entities battling over twilight islands. While this chapter charts the concept and notes some of the after effects of these encounters, it also argues for the long duration of many of these engagements. Although charting more recent histories in these locations is important, longer narratives are needed that include incremental dispossessions, ecological unfreedoms and other forms of slow violence.[4]

This chapter's focus on twilight islands does not seek to detract from the great work that has come from critics examining other types of ecosystems – environmental and otherwise. It also does not brush aside the decades of scholarly work in political economy, environmental justice and food sovereignty that has highlighted power hierarchies between entities, their access to and control of resources, or the social movements that have arisen in an effort to unite groups around common causes in response to environmental injustices or food insecurity (see Kyle Whyte, this volume). What I offer here is not so much an intervention into these activist and critical discussions, as an integration as I build on these energies to create new narratives of criticality into twilight.

I will return to this point in a moment, but I want to also stress, before going further, one additional thing that I am not doing. My framing of twilight islands does not emerge from a myopic or distorted understanding of ecology in which air, water, soil or other system dynamics are somehow bracketed and therefore contained within a geo-politically demarcated space or within the territories attributed to certain seas. I understand that ecosystems and nonhuman entities cross and enter into a range of geographical spaces. I use these bounded markers not to limit the reach and significance of twilight islands, but to work through what is undoubtedly a more global phenomenon. Let me be clear: this is not an argument for universalism, but more an awareness that the forces under consideration here are not isolated to one or two geographical zones. Twilight islands are everywhere.

Many of these twilight islands have lived contradictory lives in twilight: invisible to some; hyper-visible (and therefore controllable) to others. This dichotomy allows twilight islands to remain critically important to geo-politics while existing in the shadowed recesses of global modernities. Twilight islands in the Caribbean include those that are engineered, such as the Atlantis resort (constructed from the land and bays of the formerly privately owned Hog Island), and those ignored, such as: a) Navassa Island – (discussed above); b) Isla de Aves – a protruding rock, partially submerged during hurricanes that lies 115 miles southwest of Montserrat, but is claimed as an island dependency of Venezuela, a South American nation-state nearly 800 miles to the south; and c) Monito – an uninhabited and largely inaccessible island in the tumultuous strait between Puerto Rico and Hispaniola that remains a key geographical acquisition for the U.S.

The above set of uninhabited, barren, isolated islands are critical to the physical and conceptual geo-positioning of nations, such as the U.S.; the global economic flows of capital; and the control and management of marine and terrestrial resources within and outside the region. Some of these geo-positionings, and even islands, are part of new eco-governmental formations. Others are based on older legal or quasi-legal encounters that remain under dispute. Some of the reasons for the former disputes, such as battles over temporary sovereignty claims, have transitioned into new environmental battles in which a self-designated *planetary sovereign* – defined by Joel Wainwright and Geoff Man in an influential essay in *Antipode*, as sovereignty 'reconstituted for the purposes of planetary management' – must act at the Earth scale 'for the sake of *life on Earth*' in order to save and rescue twilight islands from humanity's destructive habits and tendencies (5).

Twilight islands may be ill served in most global environmental justice-type talks at the international and even regional levels, due to their lack of human habitation and environmental visibility. Although the eco-governance and consumptive links and networks between twilight islands may be compelling and important, the media does not cover them. Instead, attention is given to the more rapid destruction and scale of slow violence in territories such as Kiribati, in the Pacific Ocean, which has human champions who can document and call attention to the submersion of the island due to rising sea levels (see Steiner 147–180). Who is left, then, to champion twilight islands? The designated planetary sovereign(s).

Sovereignty, twilight islands, and the logics of control

Compellingly, there are untold stories here about the ways that eco-emergencies enable self-dubbed planetary sovereigns to extend their sovereign control of twilight islands. Conveniently, as in the case of Navassa Island, 'planetary sovereigns,' such as the U.S., argue that they are the only entities able – or sovereign enough – to save these islands, now coming to be known as 'biodiversity hotspots;' even though they are often the very same entities that previously sought to destroy the island's fragile ecosystems. As the above cases show, amnesia is not just about forgetting the past; it is also about ignoring the ghosts of capital and consumption that permeate our island histories. Is this amnesia intentional? Maybe. Given recent world events that saw the collapse of economies due to secretive manipulations and the movement of shadowed debt, it is not so farfetched to imagine that a select group of territories and international business conglomerates may profit from keeping knowledge about and the use of these islands in the shadows. It is hoped that this chapter offers a response to this suspicion by providing a glimpse of the understanding and knowledge gained from deploying humanistic tools, alongside others, to investigate these islands living in twilight. While this chapter has focused on a number of twilight islands, there are many more in the shadows for environmental humanists and others to now find and link to global processes of exchange that include hidden flows of politics, power and greed that align with what critical theorist and art historian Michael Betancourt describes as *agnotologic-capitalism*: 'a capitalism systematically based on the production and maintenance of ignorance' ('Immaterial value and scarcity in digital capitalism').

Of course, it is true that some twilight islands are actually known and referenced – either in scholarly sources or in governmental discussions. Amnesia is not the same as silence. Yet, while some twilight islands appear on maps and feature within frequent geographical delineations of sovereign boundaries, all exist as odd geo-political constructs. Although their continued existence is often crucial to flows of capital, ecosystems, resources, developmental aid, and human and nonhuman bodies, they nonetheless typically operate as non-functioning political sites. No representatives argue for the rights of twilight islands. No leaders represent the interests of the multiple species and matter that surround them. No political champions fight at the regional and global level for their continued survival or see the many battles over their sovereignty as part of a much larger process of political dispossession. We must use all of our resources – including the criticality of the humanities – to shine a light on these islands for the future of their regions, and the world as a whole, in order to situate them as more than just non-places.

Anthropologist Marc Augé asserts that if a place 'can be defined as relational, historical and concerned with identity, then a space which cannot be defined as relational, or historical, or concerned with identity will be a non-place' (63). Although critical of these spaces, Augé nevertheless sees them as twentieth and twenty-first century creations of supermodernity – our truckstops, airport terminals and shopping malls that slip into and out of nation-states and populate our globalised and ultra-mobilised world. I agree, in principle, with Augé, but sense that twilight islands offer distinctly different types of non-places. Instead of being involved in the physical flow of humans, twilight islands have often been set up to administratively restrict human mobility, if they are accessible at all, especially by sea.

This is the case for Monito Island. Uninhabited, inaccessible by boat and covered in sheer limestone cliffs, Monito lies between the Dominican Republic and Puerto Rico, with a land area of about 40 acres. Bombed by the U.S. military after World War II, the island is claimed by the U.S. and has been made a part of a nature reserve. Since the 1990s, efforts have been made to 'restore' biodiversity to Monito by eradicating 'foreign' species on the island. Notably, this initial poisoning campaign was stopped in 1993 due to a violation of the U.S. Federal Insecticide, Fungicide and Rodenticide Act (FIFRA) (García et al. 1–31). Follow-up assessments suggested that non-targeted species were not being harmed – so the poisoning continued for nearly six more years. While restoration projects such as the one on Monito have been attempted in many territories around the world, their use and proliferation on twilight islands by some of the same agents involved in the sovereign control (or sovereign 'protection') of these territories raises serious questions about the uses of ecological crises to extend claims to power. From being a military disposable backyard to a reserve in need of poisoning to achieve species balance, Monito has existed, as have many other twilight islands, as someone's else's environmental thing.

In his influential treatise *Discourse on Colonialism*, Martinican playwright, poet, politician and postcolonial theorist Aimé Césaire declares that the process of colonisation forces certain entities to become 'things,' consumable by others. In teasing out the ways that colonisation reenacts subjugation in everyday life as a normative act of relation, Césaire takes aim at the ways 'colonisation = *thingification*,' and destroys ecosystems in its creation of new systems of eco-governance (a condition also experienced by twilight islands):

> I hear the storm [....] I am talking about natural *economies* that have been disrupted [...] about food crops destroyed, malnutrition permanently introduced, agricultural development oriented solely toward the benefit of the metropolitan countries; about the looting of products, the looting of raw materials.
>
> (Césaire 42–43)

These colonisers, he warns, shout about progressive practices and improved conditions that have eliminated abuses. 'I too talk about abuses,' he argues, 'but what I say is that on the old ones – very real – they have superimposed others'. The past perpetrators have formed alliances with the new ones, 'to the detriment of the people'. What has emerged from this crisis is 'a circuit of mutual services and complicity' (Césaire 43). The transmutations that Césaire discusses – the old abuses/abusers and their superimposed shiny new conduits – have found new life in the crises, protection and resource allocation/extraction of twilight islands. While not 'detrimental' in intent, these new forms of services have brought together untested mechanisms that appear to work within older logics of injustice and capitalism, even as they target sustainable solutions to ecological crises.

One such example of these logics can be found within the innocuously named Caribbean Challenge Initiative (CCI). Founded in 2008, the CCI is an ambitious plan to protect at least 20 per cent of the near-shore and marine ecosystems in the Caribbean by 2020. This protection scheme has received significant global support since its inception, ensuring that interested agents would champion the cause and encourage all 37 nations in the Caribbean to take part (*Caribbean Challenge Initiative*). The governance and agreement models for rolling out the scheme have gone through a number of iterations. These models concern me – as they should any humanist or scientist interested in the distribution of power within unequal territories – but what I want to focus on here is the financing of this scheme and its scale.

In addition to the establishment of marine and near-shore protection areas, the CCI also aims to create National Conservation Trust Funds in all of its participating countries. The UK-based charity the Caribbean Biodiversity Fund (CBF) is channeling the initial funding to these Trust Funds. Set up in 2012, the CBF focuses on environmental protection and initially received global commitments of $42m – a number that is supposed to reach $200m by 2020 (*Caribbean Biodiversity Fund Fact Sheet*). Some of the CBF's biggest donors include the German government, the Nature Conservancy and the Global Environment Facility (GEF). You have probably never heard of the GEF, unless you work on climate justice issues or follow reports about the financing behind conventions, such as the United Nations Framework Convention on Climate Change. Founded in 1991, the GEF is bankrolled by a long list of stakeholders, including: the Inter-American Bank; the Food and Agriculture Organization of the United Nations; the World Bank Group; and the United Nations Development Program ('What is the GEF?'). Still following? Quick recap: the CBF will push money into the Trust Funds of the various countries in the Caribbean participating in CCI (at present count that total stands at 10, with more potentially ready to sign on). The Trust Funds are described on the CCI website as having a 'sustainable finance architecture' (*Caribbean Challenge Initiative*). Participating countries are encouraged to find additional sources of income from public and private investors interested in the financial opportunities provided by the Trust Funds.

A quick snapshot of those potential investors can be found on the list of sponsors and businesses that supported and/or attended the Phase II kickoff summit for the CCI in 2013 held in the British Virgin Islands (BVI). Sponsors included the Virgin Group (Sir Richard Branson's company), Tiffany's (the diamond variety) and the Royal Caribbean cruise line. Business attendees included Avis, Coca-Cola and American Express, in addition to other invitees, including Caribbean heads of state, other governmental representatives and various funding bodies (*Caribbean Summit of Political and Business Leaders*). Although the summit was held at two venues on Virgin Gorda (just east of Puerto Rico), guests also met on Necker Island, the privately owned island of Sir Richard Branson, who also happened to co-host the summit. The summit brochure describes Necker Island through the testimony of others who have visited Branson's home for 'environmentally themed' meetings in the past. They attest to its 'magical atmosphere' (*Caribbean Summit of Political and Business Leaders*).

Branson and his company seem strategically placed in all of the CCI material – including appearing in a short clip on the CCI website with the title 'A Commitment to the Caribbean' under the larger heading of 'The Caribbean is our home' (*Caribbean Challenge Initiative*). The Caribbean may be the home to fragile biodiversity hotspots and ecosystems in need of 'planetary sovereigns,' but it seems to also be home to a complex governance and finance scheme that compellingly involves many of the same players involved in most development and aid infrastructures: foreign governments and international corporations. A prime example? There are three co-chairs of the newly formed CCI strategic governing council. One co-chair is the government of the Bahamas. The second is the Virgin Group. The third is one of the biggest donors to the CBF, the government of Germany. I should be clear here: I am not questioning Branson's or others' commitment to the cause of environmental protection. What concerns me, as noted above, are the ways in which a select group of players may be creating new strategic economic models utilising older forms of control. The result? More protection of the Caribbean, but for whom?

Although not every twilight island in the Caribbean will be drawn into the CCI, a significant number of them will be blanketed by the 'protection' and the investment offered through the 'sustainable finance architecture' at the heart of the initiative. For this and other reasons, I have been following the CCI. Its scale remains impressive. Even as I write this, plans are being developed within the CCI for a 'private sector financial mobilisation' campaign – called Defend Paradise (*Caribbean Summit of Political and Business Leaders*). Apparently, you can save tomorrow what you contaminate today. The plans do not end there. Sustainable fishing and tourism will be key targets across the participating nations, as well as reducing or controlling marine pollution. How? Those details are still being worked through – and they are important. Yet, as I sift through the murkiness of the National Trust Funds and the playground of the new sustainable finance mechanisms, I feel as if I am re-reading a feel-good rom-com set in the Caribbean in which structural adjustment switches identities with the environment in order to win the heart of capital! Although encouraged by the interest of these various actors to find solutions to the very real environmental crises facing island regions, I must question the cost of these plans and insist that participants continue to ask themselves who, exactly, is being protected by this 'architecture'. I hope, for all of our sakes that we will not wake from our critical twilight only to ask, 'what happened not merely to the forests and the bays but to a whole people'.

Conclusion

Following the twilight that has been the focus of this chapter has taken us from islands, such as Navassa and Monito, to the sustainable finance architecture of the CCI. This move reflects the unstable nature of twilight to provoke amnesia and doubt, even as those living within its post-imperial conditions struggle to resist the spectres that remain vitally present in our ecosystems and in our politics.

By focusing on twilight islands, this chapter unveils the shifting terrain of ecological protection that appears not as a new planetary future, but as an old foe, eagerly hiding within the new networks of relationality. Although this chapter offers only a brief sketch of the form and costs of *being* in twilight for nonhuman inhabited island spaces, it has nonetheless forged a path of inquiry that breaks free of the universalising tendency within some studies to consider islands with no people as pristine natural areas devoid of human impact. Instead of tangentially in environmental and ecological crises and battles, twilight islands sit directly in the midst of them – yet, we do not see them as a set of historic spaces within which and to which old/new forms of governance and consumption run amuck.

This analysis, while novel, in parts, does speak to the call by DeLoughrey, Didur and Carrigan for an environmental humanities that is more fully integrated with political ecology, environmental justice and geography – along with postcolonial studies (DeLoughrey, Didur and Carrigan 18). 'Such perspectives,' they assert, 'are crucial to helping us think through the spatial as well as temporal interrelations between capitalism, colonialism [...and] uneven development on a variety of interlocking scales' (18). They stress that what is needed in the face of the 'rising tide of environmental catastrophes' is more examination into the 'historical and economic roots of these crises' (18). In drawing twilight islands and their histories into critical light, this chapter offers a modest attempt to remove them from obscurity and examine the roots of the 'crises' that continue to haunt these islands.

As the creation of the CCI demonstrates, the Caribbean and its twilight islands have become important players in the solutions to off-scale human alteration of the Earth's ecosystems. Yet, it and they often remain things – debated, mechanised and controlled by others. If the CCI has the potential to create new pathways of power for island peoples, it also maintains older forms of control that limit this power from being deployed. What emerges, then, is not a solution to twilight, but something that maintains the ghostly presence of past capital and the appetite of its modern day progeny. Those hungry for this capital and its related power are not prepared to be unveiled. As humanists, scientists, activists and engaged scholars, we must continue to see these deep histories and disclose them. Our future must not end in twilight.

Notes

1 Derek Walcott, 'What the Twilight Says', in Walcott, *What the Twilight Says: Essays* (London: Faber and Faber, Ltd., 1998), 4.
2 I stumbled upon accounts and details about Navassa Island while researching Haiti. I discovered that Navassa is rarely seen as part of a larger set of global 'islands with no people'.
3 See H*f*E Asia–Pacific Observatory, http://hfe-observatories.org/observatories/asia-pacific-observatory/
4 For more on these unfreedoms, see Karen N. Salt.

Works cited

Augé, Marc. *Non-places: An Introduction to an Anthropology of Supermodernity*. New York: Verso Books, 2009. Print.
Benítez-Rojo, Antonio. *The Repeating Island: The Caribbean and the Postmodern Perspective*. Trans. Jamas Maraniss. Durham: Duke University Press, 1996. Print.
Betancourt, Michael. 'Immaterial Value and Scarcity in Digital Capitalism.' *ctheory* (2010). Web.
Caribbean Biodiversity Fund Fact Sheet. Caribbean Biodiversity Fund Secretariat. N.d. Web.
Caribbean Challenge Initiative. Caribbean Challenge Initiative Secretariat. 2014. Web.
Caribbean Summit of Political and Business Leaders. CCI Summit Secretariat. N.d. Print.
Césaire, Aimé. *Discourse on Colonialism*. Trans. Joan Pinkham. New York: Monthly Review Press, 2000. Print.
DeLoughrey, Elizabeth. 'Heavy Waters: Waste and Atlantic Modernity', *PMLA* 125.3 (2010): 703–712. Print.
DeLoughrey, Elizabeth and George B. Handley. 'Introduction: Toward an Aesthetics of the Earth', *Postcolonial Ecologies: Literatures of the Environment*. Eds. Elizabeth DeLoughrey and George B. Handley. New York: Oxford University Press, 2011. 3–39. Print.
DeLoughrey, Elizabeth, Jill Didur and Anthony Carrigan. 'Introduction: A Postcolonial Environmental Humanities', *Global Ecologies and the Environmental Humanities: Postcolonial Approaches*. Eds Elizabeth DeLoughrey, Jill Didur and Anthony Carrigan. New York: Routledge, 2015. 1–32. Print.
García, M. A., R. López, F.Nuñez, J. P. Zegarra, I. Llerandi-Román, J. A. Cruz-Burgos, O. Monsegur and C. Figuerola. *Ecological Restoration of a Tropical Island: The Monito Island Experience*. U.S. Fish and Wildlife Services and Department and Natural and Environmental Resources. N.d. 1–31. Print.
Gordon, Avery F. *Ghostly Matters: Haunting and the Sociological Imagination*. Minneapolis: University of Minnesota Press, 2008. Print.

Hartman, Stephen. 'Unpacking the Black Box: The Need for Integrated Environmental Humanities (IEH)', *Future Earth* (2015). Web.

Nixon, Rob. *Slow Violence and the Environmentalism of the Poor*. Cambridge: Harvard University Press, 2013. Print.

Salt, Karen N. 'Ecological Chains of Unfreedom: Contours of Black Sovereignty in the Atlantic World', *Journal of American Studies* 49.2 (2015): 267–286. Print.

Steiner, Candice Elanna. 'A Sea of Warriors: Performing an Identity of Resilience and Empowerment in the Face of Climate Change in the Pacific', *The Contemporary Pacific* 27.1 (2015): 147–180. Print.

Sullivan, Sian. 'Banking Nature. The Spectacular Financialisation of Environmental Conservation', *Antipode* 45.1 (2013): 198–217. Print.

Wainwright, Joel and Mann, Geoff. 'Climate Leviathan', *Antipode* 45.1 (2013): 1–22. Print.

Walcott, Derek. *The Antilles: Fragments of Epic Memory (The Nobel Lecture)*. New York: Farrar, Straus and Giroux, 1992. Print.

——'What the Twilight Says', *What the Twilight Says: Essays*. London: Faber and Faber, Ltd., 1998. 3–35. Print.

'What is the GEF?' *Global Environment Facility*. Global Environment Facility, 2013. Web.

6 Seaweed, 'soul'-ar panels and other entanglements

Giovanna Di Chiro

Stories for connecting (when the world is falling apart)

Storytelling in the fields of Environmental Studies and Environmental Humanities under the sign of the 'Anthropocene' resounds with litanies of trauma, disaster and extirpation; it is at once doleful and apocalyptic. The typical Environmental Studies syllabus rehearses these familiar tropes of ruination and decline, and despite my best efforts to avert the collective numbness and the disabling environment associated with 'well-informed futility syndrome,' in my own classroom I am regularly accused of triggering despair and hopelessness. Endeavouring to address this ongoing pedagogical dilemma, in 2014 I developed a course at Swarthmore College focusing on the literary genre of life-narrative. I assigned texts comprised primarily of activist memoirs, autobiographies and memoir-esque novels. I reasoned that having students read life-writing produced by diverse scholars and activists who tell stories about living and acting in and through trauma would inspire 'hope'. The kind of hope I was aiming for, the hope one must embrace when the world is falling apart – 'hope in the dark,' as Rebecca Solnit puts it – is not about denial, delusion and crossing one's fingers while hoping for the best. Solnit explains:

> [Hope] is not the belief that everything was, is, or will be fine. The evidence is all around us of tremendous suffering and tremendous destruction. The hope I'm interested in is about broad perspectives with specific possibilities, ones that invite or demand that we act … You could call it an account of complexities and uncertainties, with openings.
>
> (12)

My hope was that through reading self-life-writing (auto-bio-graphy), a genre filled with accounts of the 'complexities and uncertainties, with openings' embodied in personal struggles and life choices, the students would gain a better sense of this meaning of *active* hope. The authors we read for the course reflect on how they come to see themselves as social and environmental change agents, having throughout their lives encountered difficulties, forged hard won lessons and patched together the critical tools enabling resilience, empowerment and creativity. I was able to get the course ('Race, Gender, Class and Environment') cross-listed in the departments of Environmental Studies, English, Sociology & Anthropology, Gender & Sexuality Studies, and Black Studies. By reading life histories of scientists, humanists and social activists, I wanted students to understand how this varied assemblage of academic disciplines was deeply and creatively entangled with modern concepts of human–nature relationships and struggles for social and environmental justice.

Activist memoir, and life-writing in general, calls on readers to explore their personal histories and trajectories, so I designed one of the course assignments with an option for students to pen a chapter of their own self-life-story. I envisioned that one of the learning outcomes of the course would be for students to imagine and give voice to the idea that they themselves were endowed with historical agency, a proactive identity that might interrupt the habitual turn to despair and cynicism. Unsurprisingly, re-reading these activist stories inspired me to examine the genealogical roots of my own 'enabling entanglements' (Tsing vii), life experiences and choices that today help me to resist the hopelessness of the Anthropocene narrative, and to imagine and lean towards creating more critical, embodied, collaborative and *life-affirming* curricula, theories and community engagements in this time of great uncertainty and in the face of a damaged Earth.

In this chapter, I deploy a mixed genre of auto-eco-biography to explore the influences and experiences of collaboration, symbioses and multi-sited nature–culture entanglements that have occupied me for the past four decades. From my membership in a 'seaweed sisterhood' while working as a marine biologist in California, Hawai'i and Washington, to my community-based action research on environmental justice and urban agriculture in Massachusetts, to forging coalitions with my students and community leaders in Philadelphia to co-produce a just transition towards solar-powered economies, I have long been engaged in multiple teaching and research muddles hoping to create alliances, loving kin and community assemblages for earthly survival in the time now being called the Anthropocene. Reading (and, perhaps, writing) activist stories and life narratives helps to move us beyond the declensionist narratives so favoured by the gloomy greens or the salvation stories championed by eco-modernists, which are two of the dominant narrative tones common to the fields of Environmental and Sustainability Studies (Di Chiro, 'Environmental Justice').

In the following notes from my own underground, background and right *now* stories as well as my teaching of this course, I remember my own involvement with a seaweed sisterhood past and with current multi-sited and multi-class, -race, -gender, and -species coalitions. Through this partial auto-eco-genealogy, I reflect on stories of the research and activism that I co-create with others, how they are comprised of multi-community constellations (human and non-human) and yet how I recognise that these coalitions will be anthropocentric, although, hopefully of an enlightened kind. A multi-species, enlightened anthropocentrism (Chakrabarty), always engaged in the arts of noticing through an environmental justice lens, recognises that while acknowledging and mourning the loss and precarity of millions of animal and plant species due to anthropogenic recklessness, some *human* populations, including many communities of colour in the U.S. living in polluted environments and facing ongoing colonialism and racial violence, profoundly and accurately consider themselves 'endangered species'. I'm interested in thinking about how my own seaweed entanglements helped me to see – and to develop the arts of noticing – the multiple, interconnected worlds comprised of different human lifeways and other species who will co-create stories of resurgence that may help us to live convivially together and to forestall, or to not go the way of, what some are predicting to be the Sixth Great Extinction on planet Earth (Kolbert).

How do we tell different kinds of stories about how we will live together, act to change the world, and build alliances for 'collaborative survival' (Tsing 28) in these times? Many activists in the climate justice movement draw on the ideas of groups such as Movement Generation and the Center for Story-based Strategy, and argue that social movement starts with your own story. Drawing on sociologist Marshall Ganz' idea of 'public narrative,' many environmental and climate justice activists are developing a strategy that 'prioritizes story and relationship, aimed at connecting people to their source of passion, to shared identity, and most of all, to hope' (Moe 49). The public narrative strategy starts with individuals telling their own stories, the story of 'Self,' and then moves to 'Us' by sharing stories of how we are entangled and what we have in common. Recognising the new dynamic that emerges in spaces when stories are shared, the story of the group shifts to co-construct a story of 'Now': the way individual and group stories become stories of a broader movement identifying what we are called to do right now and how we can take action together.

Noticing, re-membering and creating (when the world is falling apart)

Dancing on Our Turtle's Back

Stories of 'Now' told by the Indigenous and First Nations authors of several of the life narratives we read in my course embed earthwide and intergenerational trauma in critiques of ongoing settler colonialism and the advent of modern capitalism, and not in the universal humanism of the Anthropocene. In one of our course texts, *Dancing on Our Turtle's Back*, Mississauga Nishnaabeg activist and author Leanne Betasamosake Simpson argues for the need for *resurgence*: a resurgence of history, memory, and cultural knowledge to transform and 'dance a new world into being' (quoted in Klein 'Dancing the World' n. pag.). Simpson's eco-politics for addressing trauma in a damaged world express her Nishnaabeg culture's notion of sustainability: *mino bimaadiziwin* ('the good life,' or 'continuous rebirth') (Simpson *Dancing* 142). For Simpson, social movements for sustainability and environmental justice need to be 'movements to create more life, propel life, nurture life, motion, presence and emergence' (143). The practices of *creating*, argues Simpson, whether making clothes, food, shelter, stories, songs or dances were the base of Nishnaabeg society. 'Creating was regenerative, and ensured more diversity, more innovation, and more life. In essence, Indigenous societies were societies of doing: they were societies of presence' (92). In contrast, modern empire culture, framed by the logic of advanced capitalism and consumerism, she contends, requires 'both absence and wanting things in order to perpetuate itself' (92–93).

Simpson's cultural memoir weaves together stories of Self, Us *and* Now. Her book recounts stories and practices of resistance, flourishing and creativity in the face of centuries of ongoing extraction and devastation visited upon First Nations and Indigenous communities by the Canadian and U.S. governments and corporations. But, as she says, this is not the first time Indigenous peoples have 'noticed' and lived through social and ecological collapse such as, for example, 'the construction of the St. Lawrence Seaway, the extermination of the buffalo in Cree and Blackfoot territories and the extinction of salmon in Lake Ontario'. Tribal elders from all across North America, she continues, 'have been warning us about this for generations now — they saw the *unsustainability* of settler society immediately' (Klein 'Dancing the World' n. pag.). Reading Simpson's life narratives together in class, my students and I gained an understanding of her theory of *presence*, a positionality or lifeway that hones the skills of noticing the world around us. These skills include the arts of noticing the rise of global vulnerabilities and the decline of the life-supporting capacities of earth's social and ecological systems, as well as noticing and imagining possibilities for mutual transformation and regeneration. In a hopeful tone grounded in cultural history and personal experience, Simpson exclaims: 'I think that the impetus to act and to change and to transform, for me, exists whether or not this is the end of the world' (n. pag.). But, how do we talk about survival and creativity when the world is falling apart, especially in Environmental Studies courses? What stories do we notice and bring to the surface?

The sixth extinction and 1.5 million 'missing' Black men

While reading Simpson's stories of Indigenous cultural and ecological resurgence in the face of accelerating vulnerability, my students and I noticed the coincidence of two apparently unrelated news reports, both published on 20 April 2015, in the *New York Times*. In one section of the newspaper appeared the announcement that popular science-writer Elizabeth Kolbert's *The Sixth Extinction* had been awarded the 2015 Pulitzer Prize for non-fiction. Kolbert's riveting investigation chronicles the exhaustive scientific research documenting humankind's destructive impact on the Earth's life support systems and biodiversity. In great detail she outlines the evidence from around the world showing that human activities are responsible for species extinction levels so extensive that they are ushering in the 'sixth great extinction' in the planet's history.

In a different section of the newspaper, leading with the alarming title '1.5 Million Missing Black Men,' the other *New York Times* article we noticed reported on the demographic phenomenon that in U.S. cities 'for every 100 black women in the 25 to 54 age group living outside of jail, there are only 83 black men' (as compared to the gender ratio among whites being at near parity) (Wolfers et al. n. pag.). African American men are 'missing' in such high numbers, explain the authors, because they are behind bars or suffer premature deaths from homicide, police violence and higher rates of heart and respiratory disease. The juxtaposition of these two articles in a mainstream newspaper reporting on *deaths and disappearance* – one on mass extinctions of animals and plants and one on missing Black men – does not easily register as being in the same conversation about tragedy, catastrophe or extirpation. How do we notice these two apparently unrelated stories of vulnerability or precarious existence? How do we put in conversation with each other both stories of unconscionable massive extinction of animals and plant life and the recognition and assertion that Black lives *matter*. As Anna Tsing proposes, 'What if precarity is the condition of our time — or, to put it another way, what if our time is ripe for *sensing precarity*?' (20, emphasis added). Can a new social analysis based on 'thinking through precarity' in all its manifestations and unpredictable encounters help us to imagine possible connections and assemblages of humans and nature, to help make all life possible? (20).

Encountering a seaweed sisterhood

As my students and I read the activist memoirs, autobiographies and memoir-novels throughout the course, we talked a lot about 'seeing' the world through others' eyes and paying attention to how the authors notice the complex social and ecological gyres around them. The opening chapter of Ruth Ozeki's novel-memoir, *A Tale for the Time Being*, for example, focuses on a vivid scene of seaweed-filled tidepools encountered by one of the protagonists, named Ruth. While slogging along the British Columbia seashore in rubber boots, Ruth notices a sealed freezer bag entangled among the seaweed, barnacles and coastal flotsam and jetsam. Inside the freezer bag she discovers a Hello Kitty lunchbox containing a diary written by the novel's other protagonist, a troubled Japanese schoolgirl named Nao. Disentangling the plastic bag from the jumble of seaweed fronds, Ruth reads the life-story of Nao. The stories of these two far-flung protagonists, their complicated narratives, histories and the trans-oceanic gyres in which they are 'entangled,' animates the novel. Ozeki's *Tale* activated a few fertile memories that helped me reflect on the entanglements in my own research, teaching and activism. The novel's opening scene of the jumble of seaweed, oceanic debris and personal stories conjured for me a new way to imagine the interconnected, messy and even hopeful practices of working together at this time on earth, the geologic *now*. My students were amused when I told them that I had titled the first section of my eco-activist-memoir 'A Cosmopolitics of Seaweed Symbiosis'.

While myself a student of biology at the University of California, I had migrated towards the sub-field of phycology (the study of algae) mentored by Lynda Goff, at the time one of the few women scientists teaching in UC Santa Cruz' biology department. Like many students of marine biology of my generation, I had been profoundly influenced by reading Rachel Carson's hugely popular books *The Sea Around Us* (1951) and *The Edge of the Sea* (1955), which, along with *Silent Spring* (1962), I had read as a high school student. Carson's descriptions of coastal ecosystems of the Atlantic intertidal zone and the 'sense of wonder' for the ocean and all of nature, which, she maintained, must motivate environmental conservation, drew me to the wonders of marine biology. Other members of my growing pantheon of audacious women marine biologists who had confronted sexism and racism in the marine sciences in the mid-twentieth century included the 'shark lady,' Japanese-American ichthyologist and shark conservationist Eugenie Clark, whose book *Lady With a Spear* (1953) and many popular National Geographic specials on ocean conservation had enthralled me, and the oceanographer Sylvia Earle, the pioneering deep-sea explorer and now regular TED lecturer, and one of the most important scientific voices raising the alarm of the risks of ocean acidification and dying marine ecosystems.[1] Already an admirer of each of these marine scientists, I would soon learn that having chosen to study the co-called 'lower plants' in the field of phycology and later working as a seaweed biologist in the Pacific Northwest, I would be standing on the shoulders of another grandmother of marine botany, the 'first lady of seaweed,' Isabella 'Izzie' Abbott, considered the foremost expert on Central Coast Pacific algae and the co-author of *Marine Algae in California* (1992), among 8 other books and 160 scientific articles on seaweed taxonomy, ecology and ethnobotany.

Isabella Kauakea Yau Yung Aiona was born in 1919 in Hana, Maui, in the then Territory of Hawai'i, to a native Hawaiian mother and a Chinese father who had migrated to Hawai'i in the 1880s to work on the Kipahulu sugar cane plantation (Howe). Abbott would become the first native Hawaiian to earn a PhD in biology (UC Berkeley) and the first woman of colour to be promoted to full professorship in biology at Stanford University. She explains that she decided to study seaweed taxonomy and biology because, 'there were so few people doing it,' and, unlike the more formulaic biology of 'higher' plants, seaweed and algae are like trickster species: they are so biologically and taxonomically complex that they defy easy categorisation. 'Flowering plants mostly have the same kind of life history, so they become kind of boring,' Abbott contended. 'They make pretty flowers, they make nice smells; they taste good, many of them. But they're not like seaweeds' ('Long Story Short, with Leslie Wilcox n. pag.). Learning from her mother and aunties about how to identify, collect and cook native Hawaiian seaweed from Maui's intertidal zone, Abbott grew up loving local species such as Limu kala (*Sargassum echinocarpum*). Limu kala, Abbott would explain, is 'probably the most important seaweed in Hawai'i. People eat it, turtles eat it. And kala means "to forgive". It's used in purification ceremonies like ho'oponopono (the Hawaiian reconciliation process)' (Crites).

Knowledge about seaweed in Hawai'i had historically been the domain of women, passed from mother to daughter, and when Abbott became an expert not only in the local Hawaiian seaweed taxonomies but also in the Western Linnaean taxonomic system, she made it a point to demonstrate the equal importance and strengths of both systems. Abbott was professionally active well into her 80s (passing away in 2010 at the age of 91), and was an early critic of what she considered to be the questionable scientific concept of alien species and alien algae (algal species that were introduced to Hawaiian shores in the ballast and hulls of ocean liners or aircraft carriers or by scientific experiments gone awry, some of which are said to be invading and killing the coral reefs) (Helmreich). As someone who had fully embodied the naturecultures of Californian intertidal seaweed and Hawaiian Limu, she argued that her taxonomic research showed that the global distribution of algal species is much more widespread and connected by oceanic eddies and currents than scientists realise. The language of 'aliens' was ignorant and arrogant in her mind: 'it assumes phycologists already know everything about the distribution of plants and they impose their own values and ignorance onto the nature of plants themselves' (quoted in Helmreich 165). Abbott's seaweed science introduced to me the intertwined notions of seaweed culture and seaweed ecology – perhaps an incipient eco-cosmopolitical consciousness. Through the eyes of a budding scientist, this early understanding would lead me always to insist on the interdependence of knowledge systems and cultures, challenging the orthodoxies of hegemonic marine science. As an undergraduate, I enjoyed being a member of an emergent seaweed sisterhood when my professor Lynda Goff would take our lab group on field trips to Stanford University's Hopkins Marine Station in Monterey Bay, where we would meet up with Dr Abbott wearing her hip waders and carting a bucket filled with various intertidal algae specimens including *Nereocystis leutkeana* (bull kelp), which she would later bake into greenish-coloured salty sweet cakes. Several years later, while working as a seaweed aquaculture biologist in Olympia, Washington, I would again have the great honour to connect with Isabella Abbott, both of us speaking on a scientific panel on 'Pacific Seaweed Aquaculture'.[2] Her enthusiasm and knowledge about the taxonomic, ecological and cultural value of seaweed would still amaze me.

As one of the perks of being admitted into the seaweed sisterhood, I was able to work with Goff as a research assistant on Hawai'i's *Moku O Lo'e*/ Coconut Island examining the curious ecology of the then newly identified marine bluegreen alga /cyanobacterium called *Prochloron didemni*. *P. didemni* is a unicellular organism living symbiotically in the gut of a tunicate, or 'sea squirt' (*Lissoclinum patella*), a filter feeding invertebrate animal itself living symbiotically in association with the invertebrate coral 'holobiont' populating the reefs surrounding the island of Oahu.[3] The coral *holobiont* refers to a symbiotic relationship between coral species (marine invertebrates from the phylum *Cnidaria*) and unicellular algae (dinoflagellates from the genus *Symbiodinium*) and is considered one of the 'great incubators of life in the sea' (Gilbert and Epel 80). Our research team collected samples of *Prochloron* cells by in effect giving 'enemas' to tunicates: while scuba diving along the reef, we used pipettes to flush out the neon green, *Prochloron* cells from the gastric tract of the tunicate bodies. Although this may not have seemed like the dream job for an aspiring marine biologist, I was mesmerised by the multi-species worlds and symbiotic interdependencies I was witnessing. In my first marine biology research experience, studying the *Moku O Lo'e*/ Coconut Island coral *holobiont*, with its assemblages of cyanobacteria, dinoflagellates, invertebrates (coral, sea squirt) and vertebrates (fish, marine mammal and human), provided me with a potent example of 'collaborative survival': a lively multi-species entanglement now threatened by coral bleaching (the expulsion of the unicellular algal symbionts due to warming and acidifying oceans), oceanic pollution and over-harvesting of marine species.[4]

Our research team was exploring the symbiotic constellations of *P. didemni* and its reef hosts in 1979 at the height of a wave of Native Hawaiian anti-militarist and anti-imperialist struggles against the large military presence in Hawai'i (with 11 military bases and over 100 military installations on the island of Oahu alone), and campaigns to shut down several U.S. Navy installations, which had since World War II carried out military exercises including ship-to-shore and aerial bombardment, amphibious napalm and rocket training, and ordnance disposal (Ireland). At the time, we were blissfully unaware of the social activism happening nearby led by Protect Kaho'olawe 'Ohana (PKO), the local Hawaiian and transnational movement fighting to stop the Navy's bombing of the neighbouring island of Kaho'olawe, although we were not completely shielded from the military's reach. In striking distance from *Moku O Lo'e*/ Coconut Island, our exquisite, tropical research site and the idyllic setting for the TV show *Gilligan's Island*, lies Kaneohe Marine Air Base, a Marine Corps facility and air station built in 1919. While the air base was no longer used for gunnery range exercises in the 1970s, we could still hear the roar of Marine helicopters and the sonic booms from the take-offs and landings of military aircraft as we scuba-dived in the warm waters of Kaneohe Bay exploring Coconut Island's multi-coloured reef. While my reef diving expeditions allowed me to *notice* the dazzling complexity of this marine ecosystem, what I wasn't noticing was how Hawai'i's geopolitical location had made it one of the jewels of U.S. military strategy in the Asia–Pacific region, and had helped to sustain the profligate use of natural resources and fossil fuel energy – later referred to as the Anthropocene's 'Great Acceleration' – a growing military-industrial complex that would likewise exploit and colonise many islands, territories and peoples around the world. At the time, we also did not know how seriously implicated the fossil fuel-guzzling military-industrial-complex would be in the extensive coral reef bleaching caused by climate change that scientists are observing today in the ecologically vital and culturally unique Hawaiian reefs that had so intrigued me.

As a newcomer to the seaweed sisterhood, I felt a sense of excitement to be conducting field research in Professor Izzie Abbott's ancestral stomping grounds, although it was never easy to explain to people that I was a 'real' marine biologist since I was studying cyanobacteria and not humpback whales or bottlenose dolphins, the charismatic mega-fauna that are more likely to capture the public's imagination. Only later would I learn of the path-breaking work of Lynn Margulis, who in the 1990s expounded on the concept of *symbiogenesis*, a new theory of evolution that put cyanobacteria and their unicellular, prokaryotic relatives at the centre of the world-changing story of evolution and the origins of the eukaryotic cell, positing that new levels of biotic organisation occurred through clumping, joining, gathering, collecting, miscegenation and infection (rather than through random genetic mutation as the core mechanism of innovation and change) (Margulis and Sagan).

After graduating from college I worked for several years as a seaweed biologist and marine aquaculture specialist in Olympia, Washington, and conducted laboratory and field research to determine the optimal biological and environmental parameters within which local edible and commercially valuable intertidal seaweed species (such as *Porphyra yezoensis*) could be coaxed to grow happily on polyethylene nets floating on the surface waters of the southern Puget Sound. The goal of the project (funded by the Washington State Department of Natural Resources and the Federal Sea Grant programme) was to develop a small-scale aquaculture industry for the Puget Sound designed as a form of sustainable economic development that would boost the local economy and preserve the natural environment. As it happened, *P. yezoensis*, unlike its Japanese cousin *Nori*, would not easily submit to net-growing life in the Pacific Northwest bioregion for a host of biological, geographic, climatological, social, cultural and economic reasons.

The Pacific Northwest's typically rainy winters wreaked havoc on the seaweed's survivability, reducing the salinity of the Puget Sound's surface waters that bathed the sprouting algal gametophytes as they clung to the net's plastic fibres. Rogue logs and milling residue, having escaped from one of several mills owned by the Weyerhauser Corporation, would become entangled in the nets, break them loose from their cement-filled 50-gallon-drum anchors and set them adrift. Waterfront landowners complained to the Department of Natural Resources that the coastal aesthetics of the Puget Sound (and homeowners' living room vistas of the pristine coastal landscape) would be marred by the unsightly floating aquaculture nets. Local Japanese, Korean, Vietnamese and Pacific Island communities registered their preference for the colour, flavour and texture of the imported, commercially available varieties of seaweed, and often harvested their own plants. Members of the Makah, Quileute, Coast Salish and Skokomish communities were concerned with how the expansion of seaweed aquaculture farms would limit their access to the Puget Sound marine lands and endanger their sovereignty rights to fishing, collecting and foraging (Deloria; Norman). What to me had seemed a straightforward marine biological puzzle and an obvious environmental 'good' — how to successfully cultivate particular seaweed species in the Puget Sound to create sustainable development enterprises and preserve the ecological integrity of the marine lands — became intertwined with local politics, cultures, economies and ecologies.

At the time I would have never dreamt that I would soon become a card-carrying member of the *Amazon Scientists*, a spirited group of feminist scientific researchers at the University of Washington's School of Fisheries who had joined together in response to the widespread sexism they had all experienced working in the notoriously male-dominated field of marine fisheries. These scientists understood that being a woman in the sciences challenged the naturalisation and normativity of the gendered ideologies that assigned women and men to particular social roles and social locations. The Amazon Scientists deconstructed the dichotomous thinking that rendered the categories 'woman' and 'scientist' incommensurable in very tangible and painful ways that were manifested in the women's own professional lives. Despite the vibrant solidarity and critical analyses afforded me through my affiliation with the Amazon Scientists, the

Figure 6.1 The author hauling seaweed nets in Puget Sound, Olympia, Washington, 1981. (Photo by Giovanna Di Chiro)

conceptual dualisms in force in fisheries and aquaculture biology – science vs. politics, woman vs. science and nature vs. society – would become incompatible with my growing feminist–scientist sensibilities, and I left the marine laboratory to pursue what I hoped would be a more interdisciplinary path. More than three and a half decades after I stepped away from the day-to-day world of fisheries biology, scientific evidence continues to mount confirming the incompatibility of sustaining the world's marine fisheries (and the human and non-human communities whose lives are entangled with them) while embedded in the nature–culture dualisms and logics of domination of modern industrial societies: the extractivist mindset at the root of overfishing, oceanic pollution and global warming have led to devastating declines in the world's fish populations and spawning grounds, and a loss of livelihood and cultural heritage for many human coastal cultures, including native communities in Hawai'i and the Pacific Northwest (Longo et al.).

Had I still been in Olympia working as a seaweed biologist in cahoots with the Amazon Scientists, I would likely have joined up with the energetic, eco-politics of the 'kayaktavists,' a diverse coalition of climate justice activists, Indigenous leaders, Asian Pacific Island artists and ocean conservationists who came together in Seattle to paddle *en masse* to protest the environmental injustices and ecological devastation created by oil drilling in the Arctic Circle. In early 2015, Filipina-American artist-environmentalist Katrina Pestaño and Allison Akootchook Warden, a Native Alaskan Iñupiaq rapper, organised a demonstration to protest Royal Dutch Shell's offshore oil rig called the Polar Pioneer, which had set up anchor near Seattle *en route* to its oil drilling expedition in Alaska's Chukchi Sea. Offshore drilling in the Arctic Ocean, the activists argued, could threaten the lives and livelihoods of Iñupiat, G'wichin, Saami and other Indigenous peoples and the seals, polar bears, walrus and other non-human species dependent on a thriving Arctic ecosystem. (Juhasz). Like the intersectional activism supporting the co-survival of human and non-human oceanic ecologies practised by these kayaktavists, poets and singers, my own experiences as a marine biologist and my growing awareness of seaweed symbioses and cultural ecologies would spark my future and ongoing participation in nature culture engagements beyond the great blue sea.

'Nature as community': from coral reefs to urban streets

Still drawn to the ideas and methods of biology but seeking what I would later term a more 'embodied ecology,' or more engaged forms of science and environmental studies that would integrate more fully my feminist studies and interests in social justice politics, I entered a graduate program at the School of Natural Resources at the University of Michigan. Focusing on environmental education, I collaborated in the early 1980s on a school-based project in Detroit with one of my advisers, William Stapp, one of the first environmental studies scholars to imagine cities and urban landscapes as important sites for environmental education and sustainable development. Prioritising youth involvement, our urban environmental and action-based research activities included water quality monitoring and river clean-up projects, transforming vacant lots into community gardens and conducting youth-led asthma surveys in the local schools and neighbourhoods. Together with my other adviser, Bunyan Bryant, a pioneer in environmental justice scholarship, we integrated the then disparate fields of urban studies, environmental education and social justice theory co-producing curricula and educational materials focusing on action research in environmental education for the Detroit and Ann Arbor Public Schools (Bull et al.).

The opportunity to trace this eco-genealogy, along with my students, compelled me to notice and reflect on how my research and action experiences have taken me from my early affiliations with the seaweed sisterhood to my current entanglements in vibrant community-based collaboratives and environmental justice projects. The ecological knowledge about complex symbiotic, multi-species relationships in marine ecosystems that I had learned as a seaweed biologist would now inform my nature culture entanglements in urban environmental studies: working towards improving community health and environmental quality in low-income and marginalised communities, redeveloping abandoned brownfields and vacant lots to generate local businesses and build affordable housing, promoting food security and sovereignty through community gardens and urban agriculture, and working to advance a 'just transition' to community-generated, renewable energy. Although my work over the years has shifted from conversations with seaweeds and their symbionts to collaborations with urban communities and their allies, my experiences in both sites have given me a grounded education on the destructive legacy of the Western nature–culture binary so central to mainstream environmentalism and environmental studies in the U.S. One of the key theoretical reinventions of and interventions into the nature–culture binary that profoundly influenced my own praxis was the emergence in the 1980s of the theory and practice of environmental justice.

Based on the conceptual innovations of environmental justice and my developing action research work with community activists, in 1996 I wrote an essay called 'Nature as Community,' in which I argue that the growing field of Environmental Justice advocates a 'reinvention of nature' that rejects the epistemological/ontological tenet that separates the 'nonhuman natural world from nonnatural human communities' ('Nature as Community' 317). Adopting an ecological standpoint of interdependence, environmental justice activists re-integrate their environments and communities through an eco-politics of articulation. Such a re-articulation of the 'human place in nature' is a critical step in transforming Environmental Studies as we know it, but it represents only the most recent innovation in a longer history of enacted naturecultural understandings. As noted by Leanne Betasamosake Simpson in her book *Dancing on Our Turtle's Back*, which my students and I read in the course described above, many Indigenous societies possess complex environmental management systems that never imagined humans as 'other' to an exoticised 'nature,' and instead rely on ontologies of *relationality* rooted in an understanding of the interdependence among humans and the non-human world (LaDuke; Adamson). The ecological thinking and doing introduced to my class through Simpson's life-writing contributes to the production of a revitalised, *intersectional* environmental studies that possesses the capacity to resist the climate of despair, hopelessness and 'game over' cynicism so central to the discourse and politics of the Anthropocene.

Creating action research assemblages: just sustainability as creative entanglement

The insights I acquired from environmental justice activists and Indigenous writers on intersectional environmental studies and 'collaborative survival' would continue to inspire my research and teaching. As Leanne Simpson wrote, our environmental and sustainability movements must be creative, grounded in practices of *doing* that create, propel and nurture 'more life, motion, presence, and emergence' (143). With this vision in mind, soon after moving to Swarthmore College in 2012, I participated in several community events where I met a Philadelphia-based activist named '.O'. As a Black, lesbian, Quaker community activist, .O was interested in connecting her commitment to earth-based spirituality and racial and gender justice with her work to support the needs of the people living in 'Philadelphia's harshest ghetto'.[5] .O works with the local faith community on restorative justice and community building in North Philadelphia offering programmes through *Serenity House*, a community ministry centre supported by the United Methodist Church, which 'serves and provides a sanctuary for the social, spiritual, and human development needs of the predominantly low-income African American residents facing poverty, unemployment, substandard housing and gentrification, gun and police violence, failing schools, drug and alcohol dependency, and environmental illnesses'.[6]

Resonating with the environmental justice framework I shared with her, .O and I discussed how we could build a productive partnership supporting the dreams of communities in North Philadelphia. Over the next several years, students in my courses worked together with .O and other local residents from Serenity House to forge a coalition of gardeners, climate justice activists, engineers, horticulturalists, green jobs trainers and solar system installers to build two community gardens, organise gardening and solar energy workshops, lead community teach-ins on gentrification and community land trusts, install a solar panel on the roof of the Serenity House garage and regenerate a thriving community economy supporting local foods, green jobs, durable and affordable housing, and community-generated energy. This diverse campus–community assemblage of activist researchers would later call itself the Sustainable Serenity collaborative. One partner, community resident John Bowie, was particularly excited about the introduction of solar energy to North Philadelphia neighbourhoods: 'Bringing solar panels to Serenity House helps our children learn how to be on the cutting edge of this new energy revolution; it will also be a part of their future'.[7] These sentiments shared with my environmental studies students by residents of Philadelphia's most impoverished neighbourhoods express an embodied sense of 'sustainability' and 'resilience' in the face of the uncertainties of climate change and the threats of 'habitat loss' through gentrification.

Committed to keeping the 'soul' in sustainability, our collaborative initiated a project called Serenity 'Soul'-ar to implement our vision to support energy democracy and to create green jobs through community-owned and generated solar energy in North Philadelphia. In early 2015, with the support from a local green energy training centre, Serenity Soular organised a series of community-based energy conservation workshops introducing residents to the theory and practice of home weatherisation and methods for saving on energy costs. The Sustainable Serenity collaborative also organised with a nearby neighbourhood block association to apply for and win the city's 'Coolest Block' contest enabling 22 long-time homeowners to receive full energy efficiency retrofits. These home improvements for long-term, low-income residents on this block improve the durability and energy efficiency of their houses and strengthen the neighbourhood's capacity to adapt to increasing weather extremes including hotter summers and cold, damp winters. The lower energy costs also enable homeowners to afford to remain in their homes, defying the momentum of gentrification, which has displaced hundreds of North Philadelphia residents and threatens the integrity of other historically Black neighbourhoods throughout the city.[8]

Figure 6.2 Community Solar Workshop, organized by Serenity Soular, in Philadelphia, Pennsylvania (Photo by Giovanna Di Chiro)

In May 2015, the student leaders of Serenity Soular applied for and were selected as 'Solar Ambassadors' by the San Francisco-based solar development organisation RE-volv.[9] RE-volv's mission is to enable communities to invest collectively in renewable energy to propel the just transition away from fossil fuels. Through a year-long fellowship, RE-volv's Solar Ambassador Program supports campus–community initiatives training students and community partners in solar project design and financing, with the goal of installing a solar array on the community organisation's building. Having reached our fundraising goal in the spring of 2016, the Serenity Soular team contracted with a locally-owned solar company to complete the installation and selected local youth interested in gaining skills in solar technology to join our first green jobs training programme. Voicing the clear connections between the mission of Serenity Soular and the life-affirming goals of a 'just transition,' John Bowie states that this business model prioritises:

> the commitment to the well-being of people and the well-being of the environment. We insist that this business be based in the neighborhood, and owned by its workers, so that we are building, rather than extracting, wealth and knowledge within the community.[10]

Conclusion: stories for connecting (sympoiesis and hope in a troubled world)

At the end of the Spring 2016 academic year, this chapter of my eco-genealogy finishes on a high note: the Serenity Soular collaborative is on the cusp of installing its first community solar array in North Philadelphia, and student responses to the newest version of my environmental studies syllabus (using activist memoirs and autobiographies) contain fewer expressions of hopelessness about the world. Overall, I think most of my students relished the opportunity to envision self-life-storytelling as innovative and productive of 'active hope'. Sharing their stories, the students observed and created unpredictable connections among their diverse backgrounds and future paths.

In this chapter, I wrote a section of my genealogy to remember my ecological roots as a seaweed biologist in Hawai'i and Washington. It has been nearly four decades since my *Prochloron*-collecting days on the coral reefs of the Hawaiian islands,[11] and yet recently I find myself engaging in lively seaweed-cyano-bacteria-cosmopolitics with the developmental biologist, Scott Gilbert, my colleague at Swarthmore College. Gilbert's article, 'The Symbiotic Way of Life: We Have Never Been Individuals' draws upon new thinking in the biosciences that challenges the idea of the unitary, normative individual at the centre of history and biology, and instead argues that we are all part of lively, interspecies and interdependent bodies and other assemblages. The human 'life cycle is not one of a monogenetic individual. Rather, it is a holobiont life cycle — the integration of the host life cycle with its persistent symbionts ... Each of us develops as a community' (Gilbert and Epel 126–127). As Gilbert and Epel write in their widely used biology textbook, *Ecological Development Biology*, all life on earth is composed of 'mutualistic consortia,' the grand symbiotic relationships between

> bacteria and plants that make proteins possible, between bacteria and clams and between algae and corals that stabilize large marine ecosystems, and between fungi and plants that produce both the food and the oxygen that allow animal species (including humans) to exist.
>
> (82)

The engagements with the seaweed sisterhood helped me to understand my own complex entanglements in mutualistic consortia so necessary for life's flourishing. Remembering and writing this story enables me to contextualise my collaborative work creating community solar enterprises with local residents in North Philadelphia.

The final text we read for the course was *The Next American Revolution: Sustainable Activism for the 21st Century*, Grace Lee Boggs' essays on her theory of human (r)evolution and the resurgence from the ground up of her beloved city, Detroit. In the last decades of her 100-year life, which started with her work as a leader in revolutionary, socialist and racial justice movements and ended with her 'placemaking' work in Detroit 'growing souls' through urban gardens, community arts and youth-led civic engagement, Boggs had come to the realisation that in this 'exquisitely connected world, the real engine of change is never "critical mass": dramatic and systemic change always begins with "critical connections"' (50). Boggs urges young leaders to take action to create change rooted in their own places, to engage in participatory democracy respecting one another and all life on Earth. Each life-story we read during the semester, and each story that we wrote about ourselves, brought us face to face with the crushing social injustices and environmental depredations of our world, and with the people who 'stayed with the trouble' (Haraway *Staying with the Trouble*). In the life stories we read, there was no option to choose denial or cynicism. I hope that's the lesson my students learned in this course; to my knowledge, no one accused me of instilling gloom and doom.

Notes

1 TED is Technology, Entertainment and Design.

2 Panel on 'Pacific Seaweed Aquaculture' with Isabella A. Abbott; Michael S. Foster; Louise F. Eklund (6–8 March 1980). Institute of Marine Resources, Pacific Grove, California.

3 Bluegreen algae/cyanobacteria are single-celled organisms in the Kingdom Monera. They are considered the oldest of life forms that do not have a membrane-bound nucleus or other organelles. *Prochloron didemni* is considered an evolutionary link with the 'higher' plants because, unlike other cyanobacteria, it contains chlorophyll biomolecules and can fix carbon and generate oxygen through photosynthesis. As an 'endosymbiont' benefitting from the structure and nutrients provided by *Lissoclinum patella*, *P. didemni* provides the tunicate and its coral holobiont with photosynthetically

generated carbohydrate compounds and oxygen, which allow the coral bodies to flourish and to produce the calcification that builds the reef.

4 Later I would learn that, while teaching biology at University of Hawai'i in the early 1970s, my PhD adviser in History of Consciousness, Donna Haraway, had lived in a commune on *Moku O Lo'e*/ Coconut Island. Seaweed and other entanglements, indeed. See Haraway (2016).

5 Author's interview, O, January, 2014.

6 Ibid.

7 Author's interview, John Bowie, 19 April 2014.

8 'The Problems and the Promise: Gentrification in Philadelphia,' www.philly.com/philly/news/Gentrification_in_Philadelphia.html

9 RE-volv, https://re-volv.org/

10 Crowdrise, 'Jumpstart Serenity Soular' www.crowdrise.com/serenitysoular

11 The publication in 1980 of one of the first articles on *P. didemni* by two of the PIs I worked with, who acknowledged me as one of their field collectors. See Giddings et al. (1980).

Works cited

Abbott, Isabella and George Hollenberg. *Marine Algae of California*. Stanford, CA: Stanford University Press. 1992. Print.

Adamson, Joni. 'Cosmovisions: Environmental Justice, Transnational American Studies, and Indigenous Literature', *The Oxford Handbook of Ecocriticism*. Ed. Greg Garrard. New York and Oxford, UK: Oxford University Press. 2012. 172–187. Print.

Boggs, Grace Lee. *The Next American Revolution: Sustainable Activism for the 21st Century*. Berkeley: UC Press, 2011. Print.

Bull, J., M. Cromwell, J. Cwikiel, J. Guarino, R. Rathje, W. Stapp, A. Wals and M. Youngquist. *Education in Action: A Community Problem-solving Program for Schools*. Dexter, Michigan: Thomson-Shore Publishers, 1988. Print.

Carson, Rachel. *The Edge of the Sea*. New York: Oxford University Press, 1953. Print.

——*The Sea Around Us*. New York: Oxford University Press, 1951. Print.

——*Silent Spring*. Boston: Houghton Mifflin, 1962. Print.

Chakrabarty, Dipesh. 'Climate Change, Climate Justice, and the Anthropos of the Anthropocene,' Humanities Research Institute at Australian National University. 15 June 2012. Web.

Clark, Eugenie. *The Lady with a Spear*. New York: Ballantine Books, 1953. Print.

Crites, Jennifer. 'Pioneering Professor is First Lady of Limu,' *Mālamalama, The Magazine of the University of Hawai'i*. 21 October 2010. Web.

Deloria, Vine. *Indians of the Pacific Northwest: From the Coming of the White Man to the Present Day*. Golden, CO: Fulcrum Publishing, 2012. Print.

Di Chiro, Giovanna. 'Environmental Justice and the Anthropocene Meme', *Oxford Handbook on Environmental Political Theory*. Eds Teena Gabrielson, Cheryl Hall, John M. Meyer and David Schlosberg. New York: Oxford University Press, 2016. 362–381. Print.

——'Nature as Community: The Convergence of Environment and Social Justice', *Uncommon Ground: Rethinking the Human Place in Nature*. Ed. William Cronon. New York: W. W. Norton, 1996. 298–320. Print.

Giddings, Thomas H., Nancy W. Withers and L. Andrew Staehlin. 'Supramolecular Structure of Stacked and Unstacked Regions of the Photosynthetic Membranes of *Prochloron* sp., a Prokaryote', *Proc. Natl. Acad. Sci.* 77.1 (January 1980): 352–356. Print.

Gilbert, Scott and David Epel. *Ecological Developmental Biology*. Sunderland, MA: Sinauer Associates, 2015. Print.

Gilbert, Scott, Jan Sapp and Alfred I. Tauber. 'A Symbiotic View of Life: We Have Never Been Individuals', *The Quarterly Review of Biology* 87.4 (December 2012): 325–341. Print.

Haraway, Donna. *Staying with the Trouble: Making Kin in the Chthulucene*. Durham: Duke University Press, 2016. Print.

Helmreich Stefan. *Alien Ocean: Anthropological Voyages in Microbial Seas*. Berkeley: University of California Press, 2009. Print.

Howe, Kevin. 'Seaweed Lady, Isabella Abbott Dies', *Monterey Herald*. 17 November 2010. Web.

Ireland, Brian. *The US Military in Hawai'i: Colonialism, Memory and Resistance*. New York: Palgrave MacMillan, 2011. Print.

Juhasz, Antonia. 'Meet the Rappers and Kayaktivists out to Stop Shell's Giant Oil Rig,' *Rolling Stone Magazine*. 22 May 2015. Web.

Klein, Naomi. 'Dancing the World into Being: A Conversation with Idle No More's Leanne Simpson.' *Yes! Magazine*. 5 March 2013. Web.

Kolbert, Elizabeth. *The Sixth Extinction: An Unnatural History*. New York: Henry Holt and Company, 2014. Print.

LaDuke, Winona. *Recovering the Sacred: The Power of Naming and Claiming*. Cambridge, MA: South End Press, 2005. Print.

'Long Story Short, with Leslie Wilcox', guest: Isabella Abbott, PBS Hawaii. 17 June 2008. Web.

Longo, Stefano, Rebecca Clausen and Brett Clark. 2015. *The Tragedy of the Commodity: Oceans, Fisheries and Aquaculture*. New Brunswick: Rutgers University Press, 2015. Print.

Margulis, Lynn and Dorion Sagan. *Acquiring Genomes: A Theory of the Origins of Species*. Amherst, MA: Perseus Books Group, 2002. Print.

Moe, Kristen. 'Change Starts With Your Own Story', *Yes! Magazine* (August 2014): 47–50. Print.

Norman, Emma. *Governing Transboundary Waters: Canada, the United States, and Indigenous Communities*. New York: Routledge, 2014. Print.

Ozeki, Ruth. *A Tale for the Time Being*. New York: Penguin, 2013. Print.

Simpson, Leanne Betasamosake. *Dancing on Our Turtle's Back: Stories of Nishnaabeg Re-Creation, Resurgence, and a New Emergence*. New York: Arbeiter Ring Publishing, 2011. Print.

Solnit, Rebecca. *Hope in the Dark: Untold Histories, Wild Possibilities*. San Francisco: Haymarket Books. 2016. Print.

Tsing, Anna L. *The Mushroom at the End of the World: On the Possibility of Life in Capitalist Ruins*. Princeton: Princeton University Press, 2015. Print.

Wolfers, Justin, David Leonhardt and Kevin Quealy. '1.5 Million Missing Black Men', *New York Times*, 20 April 2015. Web.

7 Is it colonial déjà vu?

Indigenous peoples and climate injustice

Kyle Powys Whyte

Introduction

Indigenous peoples are emerging as among the most audible voices in the global climate justice movement. As I will show in this chapter, climate injustice is a recent episode of a cyclical history of colonialism inflicting anthropogenic (human-caused) environmental change on Indigenous peoples (Wildcat). Indigenous peoples face climate risks largely because of how colonialism, in conjunction with capitalist economics, shapes the geographic spaces they live in and their socio-economic conditions. In the U.S. settler colonial context, which I focus on in this chapter, settler colonial laws, policies and programmes are 'both' a significant factor in opening up Indigenous territories for carbon-intensive economic activities and, at the same time, a significant factor in why Indigenous peoples face heightened climate risks. Climate injustice, for Indigenous peoples, is less about the spectre of a new future and more like the experience of déjà vu.

Indigenous peoples and environmental change

'Indigenous peoples' refer to the roughly 400 million persons worldwide who, prior to a period of invasion, colonisation or settlement, exercised collective self-determination according to their own cultural (cosmological) and political systems. Indigenous peoples continue to exercise collective cultural and political self-determination today within territories in which they live as non-dominant populations in relation to nation states, such as the U.S. or New Zealand. Problematically, most people in the world assume it as fact that nation states have cultural and political primacy over Indigenous peoples (Anaya).

Consider the Indigenous people to which I belong, the Potawatomi, who have lived since time immemorial as an ecologically mobile (i.e. moved often within and across ecosystems for sustenance), culturally distinct and politically independent society that has maintained kinship and harvesting relationships with hundreds of particular plants and animals, and stewarded their habitats and cultivated crops across an area of 30 million acres in the Great Lakes region of North America. Potawatomi people organise themselves through diverse cultural and political institutions, from families to villages to winter/summer houses to ceremonies to bands to clans, each of which is designed to relate to plants, animals and ecosystems in some way.

Potawatomi society is 'multispecies' in the sense that it has its own conceptions of responsibility, agency and value for the hundreds of plants and animals that humans interact with in the Great Lakes region. Potawatomi and closely related Anishinaabe/Neshnabé (including Ojibwe and Odawa) peoples usually identified themselves in environmental terms based on where they resided (e.g. a river valley) (Secunda) and in clan animal (e.g. crane) or plant (e.g. birch) terms that referred to large kinship networks centred around those particular animals or plants (Bohaker).

Potawatomi people engage in political alliances with other groups, such as the Three Fires Confederacy involving Odawa and Ojibwe peoples. We have a legacy of trading with numerous other Indigenous peoples across the region and continent. As the result of French and British colonisation during the fur trade, starting in the 1600s, and then subsequent U.S. and Canadian settlement, which fragmented and relocated our society, we are now seven distinct Tribal nations on the supposed U.S. side; and there are several Potawatomi communities living within First Nations on the supposed Canadian side.

Today, each of these communities exercises self-determination in different ways, from having their own governments to continuing to identify distinctly as Potawatomi to practising certain ancient or more recently developed cultural practices associated with being Potawatomi or Indigenous North American. Potawatomi participate in numerous trading and business transactions, diplomatic relationships and cultural exchanges with many Indigenous peoples as well as with settler and other non-Indigenous societies. There are thousands of Indigenous peoples living in the world today who share comparable histories of continuing their self-determination in spite of invasion, colonisation or settlement, such as the Saami in the Arctic, the Maasai in Africa, the Maori in the Pacific and the Mapuche in South America.

Importantly, it is here, at the point where we are discussing what it means to be Indigenous, that environmental change and climate change come to the fore as significant topics. For many Indigenous peoples today, the concept of societies having to adapt constantly to environmental change is not new. Potawatomi peoples and the larger Anishinaabe/Neshnabé group have long traditions of cultural and political systems that are based on designing institutions that have capacities for adapting to seasonal and inter-annual change. The 'seasonal round' refers to such a cultural and political system. In the seasonal round, the purpose, organisation and size of cultural and political institutions (from ceremonies to villages to bands) changed throughout the year depending on what plants and animals needed to be harvested, monitored, stored or honoured.

According to my own knowledge, the institutions of the seasonal round are organised differently depending on what plants and animals need to be monitored, harvested, stored or honoured. For example, during sugar bush (maple syrup) harvesting (coming out of the winter), a small extended family unit might be the primary organisation; in the summer during berry harvesting and fishing seasons, larger bands are formed; during wild rice season (late summer, early autumn), larger rice camps convene as coordinated organisations of multiple families. Complex clan, gender and intergenerational norms overlay institutions such as rice camps or sugar bush camps, mediating each person's leadership authority and particular responsibilities to plants, animals and other humans.

In a seasonal round, Potawatomi peoples do 'not' have the same political and cultural institutions all year round, which is very different from how, say, U.S. governmental institutions operate, such as the Senate or Environmental Protection Agency. The expansion and contraction of institutions throughout the year attempts to order society to be as responsive as possible to environmental change in ways that respect how little humans ultimately comprehend of the dynamics of ecosystems. In the seasonal round, we have an example of anthropogenic activities that are not in themselves ignorant of the tight coupling of human cultural and political systems with ecological conditions. In fact, cultural and political institutions are designed to approximate, as best as can be known, the dynamics of changing ecological conditions.

A key point I want to highlight is that Potawatomi cultural and political systems are structured rather 'explicitly' on the concept that society must be organised to constantly adapt to environmental change. So the importance of being mindful of how to adapt is not a new one to Indigenous peoples such as the Anishinaabe and others. This shows that 'Anthropogenic' environmental change is not new as an idea nor does it date to the invention of Western machines or technologies. Potawatomi and other Anishinaabe/Neshnabé societies directly attempt to cultivate ecosystems, using the institutions and 'technologies' of the seasonal round, such as the implements and skillsets for sugar bushing, so that there would be ecological conditions characterised by sufficient abundance of plants and animals. This also shows that one way of adapting to change is to work directly with ecosystems, whether through seasonal burning, strategic planting or tapping a maple tree.

As trends in seasonal change are related to climate change, it is also true in a sense that something like what we understand today in the English language as adaptation to 'climate' change was also part of Indigenous cultural and political systems as community members keep track of, share and compare memories of seasonal change over the years. Anishinaabe/Neshnabé stories, traditions and memories can be used as decision-making processes for adapting to inter-annual change (Nelson; Davidson-Hunt and Berkes).

My observations in this section offer just the basic idea that the seasonal round is built on and emphasised institutions with capacities to adapt to environmental change and to relate to local ecological conditions. I am not making any comparative claims about the superiority of Indigenous systems and institutions for values such as sustainability or resilience. Though Indigenous peoples everywhere vary widely in their being more sedentary or more mobile, or having members living in less populated areas or large urban centres, I want to convey that Indigenous peoples generally are not surprised by the idea that their history consists of the adaptive interplay between their cultural and political systems and institutions and environmental change (Trosper; Colombi).

Settler colonialism and environmental change

For Indigenous peoples of different nations and heritages, because we often share ways of life and histories that explicitly consider adaptation to environmental change, we think very specifically about different kinds of anthropogenic environmental change. That is, human-induced alterations to the environment can range from the 'anthropogenic' change involved with cultivating landscapes in a seasonal round to those involved at earth system scales such as the massive amounts of burning fossil fuels that have been occurring through carbon-intensive economic activities. Colonialism, such as U.S. settler colonialism, can be understood as a system of domination that concerns how one society inflicts burdensome anthropogenic environmental change on another society.

More specifically, settler colonialism, in the U.S. context (but also others too), very specifically targets the ecologically mobile, adaptive systems of Indigenous peoples. Settler colonialism refers specifically to a system (or structure) of oppression by which one society settles the territories of another society. More precisely, the structure of oppression involves settler society seeking to fully establish itself in that territory according to its own cultural and political systems, which requires erasing the Indigenous population (see Lefevre for an overview of major sources). Erasure can be understood as the strategic process of instantiating cultural and political institutions that destroy the Indigenous mobile, adaptive cultural and political systems and institutions that are tightly coupled with certain ecological conditions. Consider some examples of one settler colonial strategy, that of 'containment,' which engenders cultural and political institutions designed to inhibit or 'box in' Indigenous capacities to adapt to environmental change. Strategic strategy of containment were used by the U.S. to facilitate the proliferation of extractive industries, such as coal mining and oil drilling, large-scale agriculture, deforestation and the creation of large urban areas – in short, the drivers of today's ordeal with anthropogenic climate change.

Consider some examples. The fixed rights of treaty areas and fixed jurisdictions of reservations, established during the nineteenth century, place limits on Indigenous peoples, effectively rendering them immobile. In the case of treaties, the idea that they represented 'fixed rights' is a U.S. settler interpretation, as Indigenous peoples, in some cases, understood the treaties as open to flexibility and renewal (Stark). Settlers eventually 'filled in' treaty areas and reservation areas with their own private property and government lands, which limits where and when Indigenous peoples can harvest, monitor, store and honor animals and plants; settlers then stigmatise containment through social discourses that cast reservations as bad places or that simply disappear reservations altogether as neighbouring communities; or settler discourses cast Indigenous harvesters and gatherers as violating 'the law,' among many other types of stigmatisation.

The consequences of capitalist economics, such as deforestation, water pollution, the clearing of land for large-scale agriculture and urbanisation, generate immediate disruptions on ecosystems, 'rapidly' rendering them very different from what they were like before, undermining Indigenous knowledge systems and Indigenous peoples' capacity to cultivate landscapes and adjust to environmental change. These changes 'contain' Indigenous peoples because they limit the abundance of plants and animals and the number of locations for harvesting, monitoring, storing and honouring. Many plants, animals and habitats are simply destroyed.

Boarding schools and other problematic forms of education strip Indigenous peoples of languages that express knowledge and skills related to particular ecosystems, seasonal change and knowledge. Anishinaabe/Neshnabé languages, for example, are derived from Potawatomi and other peoples' specific engagements with certain plants, animals and ecosystems and are primarily verb-based, referring very explicitly to particular practices arising from ecological contexts (Borrows). The forced adoption of English limits the range of meanings, knowledges and skillsets that Indigenous persons can draw on for sustenance. Many Indigenous peoples in the 1830s were forcibly removed from their territories altogether to take up small pieces of reservation land or private property hundreds of miles away in different ecosystems and climate regions.

Forms of recognition, such as U.S. Indian Reorganization Act of 1934, created Tribal governments and Alaska Native corporations that seek to redefine Indigenous peoples' governments as capitalist enterprises whose goal is to mimic the U.S. economy by investing in carbon-intensive economic activities, including Tribes that are heavily involved in coal-fire energy and mining. These governments inhibit Indigenous seasonal and clan-based cultural and political systems and institutions by creating a profit-dependent entity the citizens of which rely on its revenues for their well-being – which becomes defined primarily in terms of financial stability. The governments have been particularly hamstrung in their capacity to address key human rights issues, such as sexual violence (Deer).

All of these examples are cases where settler colonialism seeks to erase Indigenous peoples' adaptive capacity and self-determination by repeatedly containing them in different ways, destroying the ecological conditions that are tightly coupled with Indigenous cultural and political systems. While some Indigenous peoples manage to find ways to use forms of containment to their advantage, such as through winning court cases within the U.S. legal system that protect ecological conditions such as fish habitat for certain valuable species, settler colonial institutions tend to render Indigenous persons more susceptible to adverse health outcomes, sexual violence, loss of cultural integrity and political turbulence, among other common issues facing many Indigenous peoples today. In this sense, though Indigenous cultural and political institutions can adapt to change, what cannot be denied is that U.S. settler society has required rapid adaptations in which preventable harms become unavoidable harms. And Indigenous peoples incur these harms for the sake of facilitating U.S. settler society's instantiation as the dominant and legitimate nation and peoples within North America. We can look at this history as a cyclical history in which U.S. repeatedly instantiates settler institutions that contain Indigenous peoples, reducing their capacity to adapt to environmental change on their own terms (consensually) and without suffering preventable harms.

Climate change fits succinctly within this pattern. For this reason, many contemporary Indigenous peoples are concerned about their vulnerability, or susceptibility to be harmed, by impacts associated with the observed rise of global average temperature, or climate change. That is, they are concerned about climate risks as they are increasingly confronted by change stemming from the carbon-intensive economic activities of settler and other colonial societies. Climate change impacts can be seen through the lens of forms of containment (among other forms of domination), this time arising from settler contributions to increasing the concentration of greenhouse gases in the atmosphere.

Warming waters and receding glaciers affect the fish habitats in Indigenous territories all over the world, such as on the Pacific coast of North America where many Tribal nations harvest salmon for economic and cultural purposes (Bennett et al.). Sea level rise is pushing people living in the Village of Kivalina in Alaska, the Isle de St Charles in the Gulf of Mexico and the Carteret Atoll in Papua New Guinea to relocate (Maldonado et al.). In these cases we see both shrinking habitats and relocation occurring again. The Loita Maasai peoples in Africa face droughts that affect the rain conditions required for performing many of their ceremonies (Saitabu). Indigenous women, girls and two spirit persons in the Arctic and Great Plains regions are subject to greater sexual violence, abuse and trafficking as work camps for oil and gas extraction, such as 'fracking,' bring in male contractors to profit from the resources found within Indigenous territories (Sweet). Climate change impacts and drivers represent another form of inflicted anthropogenic environmental change.

Scientific reports confirm many of the threats just described. In 2014, the U.S. National Climate Assessment stated that Indigenous peoples face the 'loss of traditional knowledge in the face of rapidly changing ecological conditions, increased food insecurity … changing water availability, Arctic sea ice loss, permafrost thaw, and relocation from historic homeland' (Bennett et al. 2). The Intergovernmental Panel on Climate Change's Fifth Assessment Report claims Indigenous peoples face 'challenges to post-colonial power relations, cultural practices, their knowledge systems, and adaptive strategies' (Adger et al.).

Indigenous peoples' own descriptions of climate risk indicate that settler and other colonial societies are imposing rapid environmental change that generates otherwise preventable harms. The Mandaluyong Declaration quotes Miskito women in the Americas who say, in response to changing environmental conditions, that

> We now live in a hurry and daughters do not cook as grandmothers … We do not catch fish as before, do not cook as before; we cannot store food and

> seeds as before; the land no longer produces the same; small rivers are drying up … I think that along with the death of our rivers, our culture dies also …
> (300–301)

For many Indigenous peoples, these rapid changes are experienced as a continuation of settler colonialism and other forms of colonialism that they have endured for many years. For we have experienced these types of environmentally-related impacts before – from dietary change to relocation to sexual violence – though caused by different factors, such as multiple settler institutions of containment. Though institutions of containment represent just one limited example of a much more complex history with settler colonialism. Anthropogenic climate change is of a piece with forms of nonconsensual and harmful environmental change inflicted on our societies in the past. Some Indigenous peoples look at futures of rampant climate injustice as looking to the cyclical history of settler and other colonial inflictions of anthropogenic environmental change on Indigenous peoples in order to instantiate erasure.

Yet what is more insidious about climate injustice against Indigenous peoples is that the settler institutions such as those of containment, that inflicted environmental change in the past, are the same institutions that fostered carbon-intensive economic activities on Indigenous territories. That is, containment strategies, such as removal of Indigenous peoples to reservations or the forced adoption of corporate government structures, all facilitated extractive industries, deforestation and large-scale agriculture. What is more, and as I will discuss in more detail in later sections, these are the same institutions that today make it hard for many Indigenous peoples to effectively cope with climate change impacts. In this way, climate injustice against Indigenous peoples refers to the vulnerability caused by ongoing, cyclical colonialism both because institutions facilitate carbon-intensive economic activities that produce adverse impacts while at the same time interfering with Indigenous people's capacity to adapt to the adverse impacts.

The Indigenous climate justice movement and colonialism

Indigenous voices are among the most audible in the global climate justice social movement. Over 200 Indigenous delegates attended the 2015 Conference of Parties (COP) in Paris, France, where they pressed for (1) greater inclusion and leadership of Indigenous peoples at the COP and in the developing climate change plans of all nation states and (2) respect for the United Nations Declaration on the Rights of Indigenous Peoples (UNDRIP). Previously, at the Rio +20 Earth Summit in 2012, over 500 Indigenous persons gathered at a culturally significant Kari-Oca Village in Rio de Janeiro to discuss and express their concerns about sustainable development and climate change. Voices in the Indigenous climate justice movement call attention to how colonialism and capitalist economics facilitate the role of rich, industrialised countries and transnational corporations in bringing about risky climate change impacts. Many assert how climate injustice is a colonial structure of domination because Indigenous peoples are being erased.

The International Indian Treaty Council (IITC) states that

> Indigenous peoples have been severely impacted by the main cause of climate change, which is fossil fuel extraction carried out on our lands without our free prior and informed consent. That makes it essential that our rights are fully respected in this agreement and in the implementation of real solutions for the survival of our future generations
>
> (Native News)

For IITC, carbon-intensive economic activities occur non-consensually on Indigenous territories; establishing climate justice involves states and corporations coming to respect Indigenous rights to develop their own lands instead of being exploitative for the sake of extracting fossil fuels.

In North America, the climate justice movement includes declarations such as the Mystic Lake Declaration and the Inuit Petition and direct actions against fossil fuel industries organised by Idle No More, the Indigenous Environmental Network and the International Indian Treaty Council. Indigenous Peoples often call for stringent reductions in emissions of greenhouse gases than nations do. For example, the Mystic Lake Declaration claims that 'Carbon emissions for developed countries must be reduced by no less than 40%, preferably 49% below 1990 levels by 2020 and 95% by 2050' (The Mystic Lake Declaration 2).

Indigenous peoples take leadership in creating (or updating) and implementing their own cultural and political institutions for adapting to adverse climate change impacts. They have set up educational initiatives such as the Sustainable Development Institute at the College of Menominee Nation in Keshena, Wisconsin (founded in 1994), written their own programmes and policies, such as the Confederated Salish and Kootenai Tribes Climate Change Strategic Plan (in 2013) and – globally – designed their own metrics for how to assess climate change, such as the Indigenous Peoples Biocultural Climate Change Assessment Initiative and the Mauri Meter (for Maori cultural based environmental and climate change assessment) (Morgan).

North American networks such as the Indigenous People's Climate Change Working Group (started by Tribal college students), the Climate and Traditional Knowledges Workgroup, the Rising Voices of Indigenous Peoples and the First Stewards Symposium continue to bring attention to the climate risks faced by Indigenous peoples. Each network works to develop strategies for best dealing with climate risks. An important aspect of what these networks seek to do is resolve conflicts between Indigenous people and climate scientists that prevent both groups from working productively together to understand the nature of climate change and solutions for achieving climate justice (see, for example, Climate and Traditional Knowledges Workgroup).

In my experience as a Potawatomi, as well as an organiser and participant in the Indigenous climate justice movement, all the work just referenced in this section seeks to bring to the fore the idea that climate injustice occurs when settler and other colonial institutions inflict rapid environmental change on Indigenous peoples. Consider the 20 Treaty Tribes in Western Washington, which include the Nisqually Indian Tribe, the Lummi Nation and Quinault Indian Nation, who in the last 10 years have taken significant action on climate change. According to some of the scientific reports cited earlier, climate change, and other anthropogenic environmental change, is destroying salmon habitat in the Pacific Northwest region of the U.S.

The Treaty Tribes produced the *Treaty Rights at Risk* movement to address these problems with climate and environmental change 'as the U.S. settler state's failure to live up to its treaty obligations to do its part in ensuring it does not interfere with salmon and other species that are integral to Tribal cultural and political systems' (Treaty Indian Tribes in Western Washington 2). The Treaty Tribes have cultural and political systems that are designed to adapt to changes in salmon habitat since time immemorial. According to the late Billy Frank Jr. (Nisqually),

> Through the treaties we reserved that which is most important to us as a people: The right to harvest salmon in our traditional fishing areas. But today the salmon is disappearing because the federal government is failing to protect salmon habitat. Without the salmon there is no treaty right.
>
> (Treaty Indian Tribes in Western Washington 2)

The Tribes seek to remind and pressure the U.S. government to interpret the significance of the treaties as the Tribes do, which would require the U.S. to address how its own cultural and political institutions foster climate and environmental changes that degrade salmon habitat and erase the Indigenous relationship to the fish.

At the same time, the Lummi Nation, one of the same group of Treaty Tribes, has taken action to block the establishment of a coal shipment terminal and train railway near its treaty-protected sacred area of Xwe'chi'eXen. In addition to environmental protection, the Lummi reject the industrial capitalist values and colonial strategies that ignore treaties for the sake of expanding carbon-intensive industries such as coal. The Tribal chair, Tim Ballew II, claims 'We're taking a united stand against corporate interests that interfere with our treaty-protected rights … Tribes across the nation and world are facing challenges from corporations that are set on development at any cost to our communities' (Schilling n. pag.).

In this case, respect for treaty rights from the U.S. is a key climate justice issue. Simply calling for reductions in greenhouse gas emissions will not occur in time to protect salmon and other species. Importantly, the Treaty Tribes of Western Washington have articulated clearly an insidious feature of climate injustice against Indigenous peoples: disrespect for treaty rights concerns 'both' why U.S. has gotten away with establishing carbon-intensive economic activities on Indigenous territories 'and' why it is hard for the Treaty Tribes to adapt effectively to today's climate injustice ordeal. That is, the Tribes see respect for treaty rights as both stopping the continued interference with their cultural and political systems and curtailing the carbon-intensive economic activities that play a significant part in anthropogenic climate change. Climate justice is matter of breaking the cyclical history of colonial strategies that interfere with our environmental responsibilities, rights to self-determined adaption to environmental change and rights to reject industrial, capitalist and colonial values.

The Indigenous climate justice movement brings to the fore why it is hard to claim that 'at least' Indigenous peoples would be harmed less if rich, industrialised countries lowered their emissions without dealing with colonialism. For many strategies for lowering emissions impose harms themselves on Indigenous peoples if colonialism is not addressed. Beymer-Farris shows how Reducing Emissions from Deforestation and Forest Degradation in Developing Countries (REDD+), a United Nations programme promoting forest conservation, was implemented in Kenya in ways that displaced Indigenous peoples' cultivation of rice in protected forests, rendering these Indigenous communities worse off than before (Beymer-Farris and Bassett). Or in the U.S., bills for clean energy often exclude recognition for funding to Indigenous peoples for supporting clean energy and retrofitting of housing within the jurisdictions of Indigenous nations (Suagee). Or in the case of treaty rights or the Inuit petition, lowering emissions too slowly may render change at too lumbering a pace for the rapid changes Indigenous peoples are experiencing in some regions. Hence, lowering emissions without addressing colonialism can be highly problematic even if we assume that some of the types of solutions just referenced will ultimately have beneficial results on reducing concentrations of greenhouse gases in the atmosphere.

The bad luck view

While ongoing, cyclical colonialism is a major issue taken up by the Indigenous climate justice movement, it is rarely considered in the governmental and academic literatures that can be and often are used to understand Indigenous vulnerability to climate change and justice. These literatures include those specifically about Indigenous peoples and vulnerability, as well as literatures on climate justice that primarily refer to the vulnerability of global south countries (that Indigenous peoples live in). The latter literatures nonetheless discuss issues relevant to Indigenous peoples in the global north. In these views on vulnerability expressed in these literatures, Indigenous peoples are often seen as facing greater risks as a matter of happenstance or bad luck. Here, my goal is not so much a direct criticism of these literatures; rather, I aim to examine how well they are suited to some Indigenous peoples' situations given that they are often used to understand the nature of vulnerability and climate injustice. To begin with, the *Climate Change and Indigenous Peoples Backgrounder* produced by the United Nations describes Indigenous peoples as vulnerable

> owing to their dependence upon, and close relationship with the environment and its resources. Climate change exacerbates the difficulties already faced by

> vulnerable Indigenous communities, including political and economic marginalization, loss of land and resources, human rights violations, discrimination and unemployment.
>
> (UN Permanent Forum on Indigenous Issues 1)

The proceedings from the United Nations Conference on Indigenous Peoples, Marginalized Populations and Climate Change, states,

> Impacts on their territories and communities are anticipated to be both early and severe due to their location in vulnerable environments, including small islands, high altitude zones, desert margins and the circumpolar Arctic. Indeed, climate change poses a direct threat to many Indigenous and marginalized societies due to their continuing reliance upon resource-based livelihoods.
>
> (McLean, Ramos-Castillo and Rubis 5)

The U.S. Department of Interior makes a comparable claim, citing Indigenous peoples as 'heavily dependent on their natural resources for economic and cultural identity' (Secretarial Order 3289 Section 5, pg 4).

These understandings of vulnerability lean heavily on two ideas. First, for some Indigenous peoples, heightened vulnerability arises from their 'continuing' dependence on local ecosystems, which somehow suggests that Indigenous peoples are 'more' dependent on the environment than others. Moreover, this idea also suggests that it is by dint of Indigenous peoples' own choosing or simply happening to live in certain places and living in certain ways that they are at risk. In this sense, all that is going on is Indigenous peoples 'living close to the land' and climate change impacts occurring on top of this geographic and lifestyle situation.

Second, Indigenous peoples endure legacies of colonialism ranging from poverty to marginalisation. The resulting socio-economic conditions happen to not absorb or withstand climate change impacts well. That is, Indigenous peoples are not resilient to climate change impacts owing to their socio-economic conditions. For example, lack of employment and lack of strong infrastructures (e.g. buildings, roads, transportation options, etc.) are not going to protect communities well from climate change impacts such as sea level rise and severe droughts. That climate change impacts visit populations, such as Indigenous peoples, who have socio-economic problems, is also happenstance.

It is important to note that both ideas in the previous paragraph are not associated with the drivers of climate change. Drivers of climate change are the carbon-intensive economic activities that contribute to increasing concentrations of greenhouse gases in the atmosphere. Climate change impacts mix with Indigenous local lifestyles and socio-economic conditions to make the consequences of injustice more severe. Colonialism is rarely referenced in relation to local Indigenous lifestyles and socio-economic conditions beyond the obvious fact the latter is colonialism's legacy (Cameron; Haalboom and Natcher).

Ethicists and political philosophers such as Shue (1992), Gardiner (2006) and Preston (2012) have developed concepts of compound injustice or skewed vulnerability to describe the relationship between vulnerability and injustice, especially for people in the global south, but that also can be used in relation to Indigenous peoples who share comparable vulnerabilities to populations in the global south (and are also often living in global south countries). These concepts of vulnerability and justice extend the two ideas referenced above almost exactly. Preston summarises well some of the main threads of the literature on compound injustice and skewed vulnerabilities. He writes that 'Less developed nations had already lost the economic lottery through colonialism and military and economic imperialism,' which creates three kinds of vulnerability:

1 '*Geographically increased vulnerability to climate change*,' which refers to how climate change 'wreaks the greatest havoc and destruction on the lives of the global poor partly as a result of nothing more than geographical bad luck. Many of the global poor happen to live in locales that will be most susceptible to the consequences of increasing global temperatures …' (80–81).
2 '*Economically increased vulnerability to climate change*. Situated squarely on top of the geographically bad luck is the fact that persons who lack resources and economic mobility are less capable of extricating themselves from life-threatening situations' (Ibid).
3 '*Historical responsibility*. The geographic and economic vulnerability of the poor nations to rising temperatures is particularly unfortunate given their lack of historical responsibility for creating the problem in the first place'.

(Ibid)

For Preston, 'the three factors described illustrate how, through a combination of skewed vulnerabilities and skewed responsibilities, climate change appears to be particularly unfair to poorer nations' (Ibid).

In these readings/arguments about risk, vulnerability and injustice, the reason why harms are seen as bad luck is that climate 'injustice' primarily refers to the origins of climate change impacts on carbon-intensive economic activities, such as burning fossil fuels and deforestation. Yet these climate change drivers are 'not' related any further to the reasons why Indigenous peoples live locally in ways that are more sensitive to climate change impacts or endure the legacies of colonialism such as poverty and declining infrastructure. According to Preston's summary, climate change impacts occur 'on top of' the fact that Indigenous peoples continue to live off the land and continue to endure legacies of colonialism. Climate change impacts are like new problems that exacerbate old problems – and the old problems (e.g. colonialism) are themselves unrelated to climate change.

According to my analysis of these accounts from the U.S. Department of Interior to Preston's summation of the literature, 'bad luck climate injustice' against Indigenous peoples occurs when there is an accidental convergence of three histories. The first history is that of anthropogenic climate change brought about thanks to the carbon-intensive economic activities of industrialisation and capitalism. The second is the history of Indigenous local lifeways, that is, Indigenous cultural and political systems that have persisted over time in ways that turn out to be more sensitive to climate change impacts because they are 'resource-based livelihoods' (McLean, Ramos-Castillo and Rubis 5). The third is the history of entanglements with colonialism that render Indigenous peoples today living under socio-economic conditions characterised by poverty, isolation, discrimination and social invisibility.

Climate change impacts arise from the history of carbon-intensive economic activities 'as something new' that Indigenous peoples have to reckon with on top of everything else they have to deal with. The bad luck of the convergence of these three histories is 'something new'. Since climate change and colonialism are dissociated in the accounts of the bad luck view I am examining here, it is simply unfortunate that Indigenous and other populations are threatened the most. Accordingly, the main focus for efforts to establish climate justice should be on the responsibility of industrial countries and transnational corporations for their contributions to climate change without considering their continued participation in settler and other forms of colonialism. Resolving bad luck climate injustice primarily must involve reducing emissions and compensating victims for adaptation – solutions that remain silent on colonialism.

No case of bad luck: the village of Shishmaref

Thinking about climate injustice against Indigenous peoples is less about envisioning a new future and more like the experience of déjà vu. This is because climate injustice is part of a cyclical history situated within the larger struggle of anthropogenic environmental change catalysed by colonialism, industrialism and capitalism – not three unfortunately converging courses of history. Today's climate injustice ordeal reminds us of historic climate injustices that began well before the last 250 years of industrial development (Wildcat).

Consider one example in detail, that of Marino's work with the Kigiqitamiut people, who live in Shishmaref, Alaska, a small Iñupiat island community in the Bering Sea. The community's cultural and political systems are located at the central convergence of animal migration routes that support Inuit subsistence hunting. Because they live so far north, to most people in the world, Shishmaref is largely unknown or considered somehow spatially separate from all other societies. For this reason, the climate injustice being experienced by community members is also largely unacknowledged by the rest of the world. A community member, Fred Eningowuk, cited by Marino, says, 'Shishmaref is in the middle of a circle of subsistence' (377). Impacts such as increases in windiness, storminess and erosion and diminished sea ice threaten the low-lying island with habitual flooding and eroding Oceanside bluffs, risking a life-threatening disaster (Marino).

Globally, these changes are associated with anthropogenic climate change. But locally, Shishmaref is grappling with the immediate problem of not having the required mobility to shift and adapt in response to these impacts. Prior to settler colonisation, the community was quite mobile, with multiple adaptive institutions as part of their cultural and political systems. As Marino claims, 'Previous flexibility to environmental shifts and unexpected hazards allowed the community to adapt to abrupt changes' (374). Yet now a 'relatively immobile infrastructure and development requires people to stay in place in order to carry out their daily lives' (Ibid.). The community's immobility – a key reason for why they are vulnerable – results from colonial strategies that sought to missionise, educate and render sedentary Indigenous peoples in the Arctic, replacing the Indigenous institutions with settler ones. These policies were likely pursued out of diverse motivations, but they subsequently facilitated resource extractive industries, from fishing to the oil industry.

For example, at the turn of the nineteenth and twentieth centuries, one strategy of settler colonialism was to consolidate mobile family groups to sedentary villages with central nodes, such as a post office, government school and a mission. The last 100 years, then, is one in which the

> previously mobile Kigiqitamiut have become intimately tied to this infrastructure for school, work, life, and livelihood. Development and 'sedentarization' policies rapidly decreased the ease of mobility ... Lives and work became rooted in specific, new, critical infrastructure. Thus, traditional strategies became less practical.
>
> (Marino 378)

In light of this history, Marino argues that the village is particularly vulnerable to climate change owing to a number of key factors: (1) procedural injustice, or a lack of input into development decisions, which led to the building of infrastructure in already marginal and increasingly exposed locations (such as flood prone areas); (2) containment, or the ending of high mobility as an adaptive strategy (via sedentarisation) 'without replacing it with other readily identifiable adaptation strategies;' and (3) settler centralisation, that is, decision-making and political power were shifted outside the local community to the state of Alaska, exposing the village to distant political and economic fluctuations. 'In Shishmaref, the colonial history and inequities inherent in colonizing decisions contributes heavily to current vulnerabilities to flooding, in the ways we observe here, among others' (Marino 379).

I interpret Marino's work as suggesting three different strategies of settler colonialism to erase the way of life of Kigiqitamiut as a way of creating a U.S. homeland based on carbon-intensive economic activities. Each strategy curtails the community's capacity to adapt to environmental change. Today, we see the cycle again. Relocation is very expensive, and the policies governing relocation are highly problematic. The Federal Emergency Management Agency, who would share responsibility for the relocation, is governed by the 1988 Stafford

Act, which requires rebuilding 'in place' and 'without improvement' (378) as the way to respond to disasters. This fails to adequately protect the community. Moreover, since there is no agency that works with communities on 'preemptive disaster planning or risk reduction in these cases where erosion increases exposure to flooding hazards,' the solution requires herding the cats that are multiple federal agencies and their budgets (Ibid.).

According to Marino, the U.S. interferes with Shishmaref's capacity to adapt to environmental change, and has done so multiple times throughout its settler colonial history. The settler colonial strategies that impede adaptation today are the ones that were originally designed to facilitate carbon-intensive economic activities in the Arctic. Again, as with the Potawatomi, these recountings of cyclical history in relation to climate change are not unique to Indigenous peoples – though there will certainly always be exceptions given the diversity of Indigenous experiences globally. This cyclical history locates colonialism at the heart of the problem of both vulnerability and climate change mitigation. There is no bad luck. Climate injustice against Indigenous peoples, then, refers to the vulnerability caused by settler and other forms of colonialism 'both' because colonial institutions facilitate carbon-intensive economic activities that produce adverse impacts while at the same time interfering with Indigenous peoples' capacity to adapt to the adverse impacts.

Conclusion: the experience of déjà vu

The recounting of Indigenous histories, such as the village of Shishmaref, suggest different perspectives on the history of climate injustice than the bad luck view, which compounds three histories, one associated with carbon-intensive economic activities driven by industrial capitalism; another with colonialism and the socio-economic conditions Indigenous peoples face; and yet another with Indigenous cultural and political systems. For many Indigenous peoples, climate injustice does not involve, simply, an 'age of the human' dated to industrial development. Indigenous peoples often see themselves as participating in cultural and political systems that, from hundreds even thousands of years of experience, are explicitly designed to adapt to environmental change; climate injustice emerges as an issue more recently that is part of a cyclical history of disruptive anthropogenic environmental change caused by settler and other colonial institutions that paved the way for extractive industries and deforestation. Colonial institutional strategies that historically made it harder for Indigenous groups to adapt to climate change from the 1500s to the mid-1800s continue to complicate abilities to adapt to accelerating climate change today. We will understand the nature of climate injustice against Indigenous peoples better – and perhaps its solutions too – the more we see it as more like the experience of déjà vu.

Works cited

Adger, W.N and Juan Pulhin, Jon Barnett, Geoffrey Dabelko, Grete Hovelsrud, Marc Levy, Ursula Oswald Spring and Coleen Vogel. 'Human Security', *Climate Change 2014: Impacts, Adaptation, and Vulnerability. Part A: Global and Sectoral Aspects. Contribution of Working Group Ii to the Fifth Assessment Report of the Intergovernmental Panel on Climate Change*. Eds. C.B. Field et al. Cambridge, UK: Cambridge University Press, 2014. 755–791. Print.

Anaya, James. *Indigenous Peoples in International Law*. New York, NY: Oxford University Press, 2004. Print.

Bennett, T.M. Bull, Nancy Maynard, Patricia Cochran, Robert Gough, Kathy Lynn, Julie Maldonado, Garrit Voggesser and Susan Wotkyns. 'Indigenous Peoples, Lands, and Resources.' *Climate Change Impacts in the United States: The Third National Climate Assessment*. Eds J. M.Melillo, Terese (T.C.) Richmond and G. W. Yohe. Washington, DC, USA: U.S. Global Change Research Program, 2014. 297–313. Web.

Beymer-Farris, Betsy A. and Thomas J. Bassett. 'The Redd Menace: Resurgent Protectionism in Tanzania's Mangrove Forests', *Global Environmental Change* 22.2 (2012): 332–341. Print.

Bohaker, Heidi. 'Reading Anishinaabe Identities: Meaning and Metaphor in Nindoodem Pictographs', *Ethnohistory* 57.1 (2010): 11–33. Print.

Borrows, John. *Recovering Canada: The Resurgence of Indigenous Law*. Toronto, Canada: University of Toronto Press, 2002. Print.

Cameron, Emilie S. 'Securing Indigenous Politics: A Critique of the Vulnerability and Adaptation Approach to the Human Dimensions of Climate Change in the Canadian Arctic', *Global Environmental Change* 22.1 (2012): 103–114. Print.

Climate and Traditional Knowledges Workgroup. *Guidelines for Considering Traditional Knowledges in Climate Change Initiatives*. 2014. Print.

Colombi, Benedict J. 'Salmon and the Adaptive Capacity of Nimiipuu (Nez Perce) Culture to Cope with Change', *The American Indian Quarterly* 36.1 (2012): 75–97. Print.

Davidson-Hunt, Iain and Fikret Berkes. 'Learning as You Journey: Anishinaabe Perception of Social-Ecological Environments and Adaptive Learning', *Ecology and Society* 8.1 (2003): 5. Print.

Deer, Sarah. 'Sovereignty of the Soul: Exploring the Intersection of Rape Law Reform and Federal Indian Law', *Suffolk UL Rev*. 38 (2004): 455. Print.

Gardiner, Stephen M. 'A Perfect Moral Storm: Climate Change, Intergenerational Ethics and the Problem of Moral Corruption', *Environmental Values* (2006): 397–413. Print.

Haalboom, Bethany, and David C. Natcher. 'The Power and Peril of "Vulnerability": Approaching Community Labels with Caution in Climate Change Research', *Arctic* 65.3 (2012): 245–366. Print.

Lefevre, Tate A. 'Settler Colonialism', *Oxford Bibliographies in Anthropology*. Ed. John Jackson. Oxford, UK: Oxford University Press, 2015. 1–26. Print.

Maldonado, Julie Koppel, Christine Shearer, Robin Bronen, Kristina Peterson and Heather Lazrus. 'The Impact of Climate Change on Tribal Communities in the US: Displacement, Relocation, and Human Rights', *Climatic Change* 120.3 (2013): 601–614. Print.

Mandaluyong Declaration. 'Mandaluyong Declaration of the Global Conference on Indigenous Women, Climate Change and Redd Plus', *Indigenous Women, Climate*

Change & Forests. Ed. Tebtebba. Baguio City, Philippines: Tebtebba Foundation, 2011. Print.

Marino, Elizabeth. 'The Long History of Environmental Migration: Assessing Vulnerability Construction and Obstacles to Successful Relocation in Shishmaref, Alaska', *Global Environmental Change* 22.2 (2012): 374–381. Print.

McLean, Kirsty Galloway, Ameyali Ramos-Castillo and Jennifer Rubis. *Indigenous Peoples, Marginalized Populations and Climate Change: Vulnerability, Adaptation and Traditional Knowledge Expert Workshop on Indigenous Peoples, Marginalized Populations and Climate Change*. 2011. United Nations University. 1–45. Print.

Morgan, Te Kipa Kepa Brian. 'Mauri Meter.' 2016. Web.

The Mystic Lake Declaration. 2009. From the Native Peoples Native Homelands Climate Change Workshop II. Web.

Native News. 'For Indigenous Peoples at Cop21, the Struggle Continues', *Native News Online.Net*. 9 December 2015. Print.

Nelson, Melissa. 'The Hydromythology of the Anishinaabeg', *Centering Anishinaabeg Studies: Understanding the World through Stories*. Eds Jill Doerfler, Niigaanwewidam James Sinclair and Heidi Kiiwetinepinesiik Stark. East Lansing, MI, USA: MSU Press, 2013. 213–233. Print.

Preston, Christopher J. 'Solar Radiation Management and Vulnerable Populations: The Moral Deficit and Its Prospects', *Engineering the Climate: The Ethics of Solar Radiation Management*. Ed. Christopher J. Preston. Lanham, Md.: Lexington Books, 2012. 77–95. Print.

Saitabu, Henri Ole. 'Impacts of Climate Change on the Livelihoods of Loita Maasai Pastoral Community and Related Indigenous Knowledge on Adaptation and Mitigation', *Indigenous Peoples, Marginalized Populations and Climate Change: Vulnerability, Adaptation and Traditional Knowledge* Eds Kirsty Galloway McLean, Ameyali Ramos-Castillo and Jennifer Rubis. Expert Workshop on Indigenous Peoples, Marginalized Populations and Climate Change. Mexico City, Mexico: United Nations University, 2011. 18–21. Print.

Schilling, Vincent. 'Lummi Tribal Leaders Rally in D.C. Against Nation's Largest Coal Terminal', *Indian Country Today*. 12 November 2015. Web.

Secunda, Ben W. 'To Cede or Seed? Risk and Identity among the Woodland Potawatomi During the Removal Period', *Midcontinental Journal of Archaeology* 31.1 (2006): 57–88. Print.

Shue, Henry. 'The Unavoidability of Justice', *The International Politics of the Environment: Actors, Interests and Institutions*. Eds A. Hurrell and B. Kingsbury. Oxford, UK: Clarendon Press, 1992. 373–397. Print.

Stark, Heidi Kiiwetinepinesiik. 'Respect, Responsibility, and Renewal: The Foundations of Anishinaabe Treaty Making with the United States and Canada', *American Indian Culture and Research Journal* 34.2 (2010): 145–164. Print.

Suagee, Dean. 'Tribal Sovereignty and the Green Energy Revolution', *Indian Country Today* (19 September 2009). Print.

Sweet, Victoria. 'Rising Waters, Rising Threats: The Human Trafficking of Indigenous Women in the Circumpolar Region of the United States and Canada', *MSU Legal Studies Research Paper* 12–01 (2014). Print.

Treaty Indian Tribes in Western Washington. *Treaty Rights at Risk: Ongoing Habitat Loss, the Decline of the Salmon Resource, and Recommendations for* Change, 2011. Web.

Trosper, Ronald L. 'Northwest Coast Indigenous Institutions That Supported Resilience and Sustainability', *Ecological Economics* 41 (2002): 329–344. Print.

UN Permanent Forum on Indigenous Issues. *Climate Change and Indigenous Peoples: Backrounder*. New York, NY: United Nations, 2008. Print.

United States Department of the Interior, *Secretarial Order* 3289, 2009. Web.

Wildcat, Daniel R. *Red Alert! Saving the Planet with Indigenous Knowledge*. Golden, CO, USA: Fulcrum, 2009. Print.

8 Gathering the desert in an urban lab

Designing the citizen humanities

Joni Adamson

> for humans to survive in the long run we must become more attentive of, and connected to, the planetmates that support us within the long-lived global ecosystem.
>
> Dorion Sagan, 'Evolution', *Keywords for Environmental Studies*

Creosote, a scrubby desert bush that scientists refer to as *Larrea tridenta*, has figured prominently in my own teaching and research for many years. I first learned about this plant and the stories the Indigenous O'odham of North America tell about it in the late 1990s. The tiny, resinous leaves of creosote smell like rain when they are crushed in your fingers and always remind me of the years I was driving back and forth between Tucson, Arizona, and Sells, Arizona, the capital of the Tohono O'odham Nation. I was teaching college-prep composition and literature courses, sponsored by the University of Arizona, for O'odham high students. To get to the high school, located close to the US–Mexico border, I would drive through miles of creosote and other desert plants. I write about these experiences in *American Indian Literature, Environmental Justice, and Ecocriticism* (3–14).

The O'odham, meaning 'the People,' include several distinct bands descended from the ancient Hohokam. Today, the O'odham continue to rely on their extensive ethnoscientific and agroecological knowledge, which is passed down through stories about creosote and other plants and animals of the region that have been told for generations. This knowledge has given them the ability to survive in the Sonoran Desert, one of the driest regions on earth, for at least 10,000 years.

'Hohokam' means 'those who have gone' in the Uto-Aztecan language spoken by the Akimel O'odham, or River People, one of the bands of O'odham who still live in the Phoenix River Basin today. Phoenix, Arizona, a megacity of four and a half million people, minimal rainfall, scorching heat and an insatiable appetite for unrestrained growth, sprawls around the ancient ruins of a Hohokam observatory called 'Pueblo Grande' or the Big House, a structure located amidst other smaller buildings.[1] The Hohokam were highly astronomically sophisticated, thriving from 450 BCE to 1450 CE, a period during which they engineered observatories throughout the region to watch the stars and gauge the seasons.

Archeologists have uncovered rooms in Pueblo Grande where a shaft of light stretches from one doorway to another, signalling the exact sunrise of the summer and winter solstices.[2]

This structure, built to gather astronomical knowledge, can be linked to other civilisations around the world, from the Maya, to the Chinese, to the Austronesian paddlers guiding their canoes from 'star charts,' that are figuring in 'conversations' about the Anthropocene urged by Linda Hogan and Deborah Bird Rose in this volume. Indeed, the Hohokam contribute to this widening conversation because of their amazing architectural and engineering feats. By 1100 CE, they had built a complex system of irrigation canals to support extensive agricultural activities that supplied the approximately 40,000 people settled in the River Basin. This civilisation also supplied the travellers constantly moving back and forth between Aztecan city-states in what is now called Mexico and the Puebloan cities located in what is now called the 'Four Corners' region of North America. As explained in the Introduction to this volume, in creating a globally networked system of observatories, Humanities for the Environment (H*f*E) researchers have been inspired both by ancient places of learning and civilisation such as Pueblo Grande and by modern international scientific initiatives such as the United Nation's research platform, Future Earth. The goal is to create humanities 'laboratories' or 'research spaces' where questions about linked social and environmental changes can be considered (Adamson, Chapter 1).

H*f*E researchers have also been inspired by long-lived species such as creosote that become 'seeing instruments' that help humans consider deep time and evolutionary adaptation. In my work, I have used the term 'seeing instrument,' long associated with observatories, astronomy and cosmological story cycles that emerged from ancient centres of astronomy, as a helpful neologism for thinking about contemporary novels, poetry, street theatre, art, time-lapsed photography and blockbuster films that seek to make the complex connections between biogeochemical processes and the life on the planet more visible to the general public (Adamson *American Indian* 145; Adamson 'Humanities' 136).

Most people travelling by car between Phoenix, Arizona and Los Angeles, California pay little attention to the miles of creosote on either side of the US Interstate and know little if anything about this living 'seeing instrument' or the sophisticated Hohokam civilisation that thrived in the Sonoran Desert alongside this plant. They may be quite unaware that creosote has been continuously growing for thousands of years. The plant can be dated back to a time even before the Hohokam began developing their remarkable social, cultural, astronomical, agricultural and engineering activities. Indeed, the roots of creosote grow in clonal circles, making what may seem like many plants, one large individual. Some of these individuals have been alive for up to 12,000 years. In *The Oldest Living Things in the World*, fine arts photographer Rachel Sussman's images of creosote become seeing instruments as they capture both the robustness and fragility of a plant with a life span that dwarfs the longevity and history of the nation in which it grows, the United States (Sussman, *The Oldest*, 18–22). In a highly-viewed TED Talk, Sussman frames her photographs of creosote and other long-lived plants as an invitation to step outside of 'quotidian existence and consider a different time scale' that includes both persistent (slow) evolution and (rapid) environmental change as greenhouse gases warm the planet (Sussman, 'The World's').

In this chapter, using my research on Pueblo Grande and creosote, analysis of Ofelia Zepeda's poetry about O'odham life and Rachel Sussman's photography of ancient plants, I discuss an applied humanities research project created by H*f*E researchers networked through the North American Observatory (NAO).[3] Called 'Life Overlooked,' this pedagogical experiment links courses taught by Stephanie LeMenager at the University of Oregon, Catriona Sandilands at York University and Joni Adamson at Arizona State University. LeMenager designed a template syllabus and taught a pilot version of the course after which Adamson and Sandilands adapted the syllabus for use at their respective universities. While elements of each version of the course were different, students at each university were given a common digital assignment. Each student created a multimodal narrative or web page that was uploaded to Facebook, Weebly or the H*f*E international website. Each narrative focused on one 'non-charismatic urban species' that could be easily researched by the student at her or his campus because it lived on the campus or within walking distance of the campus. The goal was to enrich understanding of 'citizen science' by empowering students to become what project designers termed 'citizen humanists' tasked with disseminating local ecological knowledge, and rethinking time and evolution, in new artistic and digital ways that might be made accessible to widening publics increasingly interested in understanding changing biogeochemical processes at both the small and large scales.[4]

My own version of the course was designed to pilot 'new constellations' of humanities and scientific practice modelled on the narrative and research methodologies of Gary Nabhan (as illustrated in his collection of essays about Sonoran and Mohave desert plants *Gathering the Desert*) and Lyanda Lynn Haupt (as illustrated in her collection of essays about common 'backyard species' *The Urban Bestiary*) (Haupt 8).[5] My aim for the course was to teach students to write compelling narratives that might be used to reach communities of readers, government officials and policy-makers interested in forging a 'future we want'. In the spirit of creating new humanities research spaces, I taught my course at two of Arizona State University's four campuses that have been purposely designed as urban 'laboratories' where research on sustainable cities is a priority. Each of these campuses is located near ancient Hohokam observatories that act as seeing instruments as they continue to mark the annual solstices in light and shadow twice a year.

From local pedagogy to urban lab

As Dorion Sagan, quoted in the epigraph above has observed, 'for humans to survive in the long run we must become more attentive of, and connected to, the planetmates that support us' (Sagan 116). But, as Humanities for the Environment

(H*f*E) researchers ask in their 'Manifesto for Research and Action,' can humans become less anthropocentric and more attentive to their planetmates? Can they 'change tack in a more proactive way not just when confronted with immediate disaster but also when faced with a long-term threat?' (Holm et al. 984). In the Anthropocene, it seems clear that the question is not whether or if humans can or should respond, but how quickly 'and how successfully?' (Holm et al. 984). More specifically – for megacities such as Phoenix – can humans learn from past civilisations or 'from past challenges of resource scarcity?' (Ibid).

These are important questions to be asking in a place that has been dubbed by cultural critic Andrew Ross, in the subtitle of his book, *Bird on Fire, 'the World's Least Sustainable City*'. In ecological circles, Phoenix has a rather bad reputation because of the scale of the challenges it faces in a time of accelerating environmental change and widening social disparities. However, Ross argues, paradoxically, that because of its challenges Phoenix presents one of the best locations in the world for exploring human motivations and agency and experimenting with pathways that might lead to more desirable, sustainable futures. In the US, 80 per cent of the population lives in urban areas and Phoenix is experiencing some of the most rapid population growth in the nation (Ross; Anderson and Woolsey). Current migration trends suggest that urban desert populations will continue to grow and nonhuman species living in urban places will face increased extinction rates (indigenous birds) or increased opportunities to thrive (starlings, coyotes, raccoons) as temperatures rise. Ross credits ASU with systematically mapping the region's environmental features, and creating the Global Institute of Sustainability (GIOS) to marshal the university's resources towards finding 'planetary-scale solutions for the ecological crisis' not only for the benefit of the local population but for the benefit of residents of cities worldwide (Ross 176). With over 80,000 students, ASU has purposely redesigned itself over the last 15 years as an 'urban laboratory' where researchers and community leaders are seeking large- and small-scale solutions to water, food, energy, and transportation challenges.

The course I adapted for the Life Overlooked project, 'Environmental Creative Nonfiction,' was one I developed while teaching on a small campus only five miles as the crow flies from the US–Mexico border in the late 1990s. Then, there was widespread faith among environmental literary critics that creative nonfiction focused on wilderness or threatened local places would motivate people to care more deeply about their own environments. After taking my first job, I began testing my own theories about a 'local pedagogy' that might catalyse 'transformative social change' by developing curricula that was well informed by 'understanding of local people, cultures, histories, and geographies' (Adamson *American Indian* 93). My campus was so rural that my students and I could walk to nearby sandy trails that wound through stands of creosote, mesquite trees and prickly pear cactus. I always began the course with sections of Henry David Thoreau's 'Walking' and in the time-honoured tradition of the naturalist, my students would bring their nature writing journals and we would stop for writing breaks as we walked. Our readings ranged through travel narratives, memoir and nonfiction, but usually always with a focus on the desert. We read Álvar Núñez Cabeza de Vaca's memoir *Castaways*,

O'odham poet Ofelia Zepeda's chapbook *Ocean Power* and Tucson-based writer Patricia Preciado Martin's memior *Songs My Mother Sang to Me*. We would discuss the virtues of Thoreau's notions of gaining knowledge through walking and 'peripatetic' observation of the natural world.[6] We also considered how writing creatively, about environmental subjects, required gathering, then arranging and nurturing both cultural and scientific knowledges, just as Gary Nabhan so skillfully illustrates in his collection of essays, *Gathering the Desert*, about Sonoran Desert plants and animals.

I designed the course to be part literary analysis and part creative writing workshop. My students and I focused broadly not only on walking as a metaphor for writing, but on the ways that 'nature' involves not stasis, but the processes of seeing, moving, walking, migrating and mobility. For thousands of years, I told my students, humans, animals, seeds, foods and plants have walked, or been shipwrecked, carried, blown, sown or escaped into new places around the globe. For example, we would read about Spanish conquistador-turned-healer Cabeza de Vaca, who travelled through the Sonoran Desert region at the end of the sixteenth century where he saw examples of the 'Big Houses' or 'Pueblo Grande' architecture built by the Puebloan and Hohokam peoples. Cabeza de Vaca was shipwrecked in Florida, then wandered among the Indigenous peoples of North America for nine years before finding his way back to the Spanish in Mexico City. Before arriving there, he travelled through what is today El Paso, New Mexico, in January 1536, then travelled up the Gila River, one of the key waterways of the Hohokam/O'odham. Arriving in the region of modern-day Phoenix, he learned that he would need to turn south to find the Spanish and travelled along the Santa Cruz River through what is today Tucson, into Mexico. He followed routes long traversed by Indigenous North American peoples moving among the ancient villages found throughout the Sonoran Desert.[7] My students came to see how Cabeza de Vaca's first language, Spanish, failed him when he encountered the tactile and nutritional qualities of prickly pear for the first time. After leaving Florida, at the beginning of his journey, he wandered cold and naked, having fallen on rather hard times after a shipwreck. He depended completely on the peoples who knew the foods of North America best, and who helped him survive by introducing him to a plant he envisaged within Old World categories of known fruits and plants as an oddly shaped thorny 'apple-like' fruit. Today, in our entangled representations of species, we know this fruit as the prickly pear. For pedagogical purposes, Cabeza de Vaca's awkward descriptions of a plant he had never known before allowed me to teach my students how we come to 'see' plants, animals and insects we may have never noticed or are coming to know for the first time. In the adaption of the course for Life Overlooked, this passage in Cabeza de Vaca's memoir would be crucial to my students' thinking about why it is that some species are overlooked and the resulting consequences for species that humans overlook.

A little more than 300 years later, in the late 1870s, the first Mormons would arrive in the region previously traversed by Cabeza de Vaca. Later, in the early twentieth century, they were joined by other Anglo-Americans. These new settler-colonials dug out and re-engineered the ancient Hohokam canal system as a way of controlling and directing the water in the large rivers, the Gila and the Salt, that once flowed year round towards the Colorado River on the Arizona-California border. The new settlers imagined themselves to be building a city they would call 'Phoenix' after the mythical 'bird on fire' arising from the ruins (Ross). By the 1950s, the once wide-running rivers along which the Hohokam and other Indigenous North American peoples had travelled by canoe were becoming dry, due to increased agricultural use, damming upstream and pumping of aquifers to satisfy the need for water for swamp coolers and other modern conveniences that were making settler colonial life in the desert possible for non-Indigenous newcomers (Adamson *American Indian* 3–14). The O'odham were no longer able to farm along the Gila and Salt Rivers because the water table sank so dramatically that the water dried up.

This is the situation that sets the scenes for the poems in Ofelia Zepeda's collection of poetry, *Ocean Power*, which my students read after finishing *Castaways*. In the late 1800s, Zepeda's people, the O'odham, were forced onto reservations, and in the early 1900s, into the US wage economy. They began working as cotton pickers for Anglo-American farmers in the Phoenix River Valley. Zepeda grew up in fields where she was expected to thin and pick cotton. In 'The Man Who Drowned in an Irrigation Ditch,' she writes, 'We are called "the people of the cotton fields"' (Zepeda 30). As migrant workers, they were exposed to many risks, including pesticide poisoning and farm accidents. Zepeda remembers one summer evening when the women were called out to an irrigation ditch. A well-loved elder man had slipped and fallen into the water and drowned. When the women reach the ditch, they respond in unison. 'With a single vocal act they release from the depths a hard, deep, mournful wail. / This sound breaks the wave of bright summer light above the green cotton fields' (33).

Zepeda connects stories about the traditional migration routes that took the O'odham through the desert to collect wild plants and foods, and to the Pacific Ocean to collect salt, which inspires the title of the collection, *Ocean Power*. The poems also discuss traditional medicines including creosote. Writing about this plant, ethnobotonist Gary Nabhan calls creosote the desert's 'drugstore' because the O'odham have uses for the plant that extend from birth to death (11–19). A tea made from the plant is used for colds, chest infections, intestinal discomfort, nausea and swollen limbs, and to ease the pain of childbirth (Nabhan 14–15). The plant exudes droplets of a pungent, rain-like smelling syrup or resin on its stems. Zepeda writes that her mother is buried with creosote lining her grave, so she can have 'the smell of the desert with her, / to remind her of home one last time' (40).

Like other Indigenous peoples around the world, the O'odham have begun reclaiming their ancient medicines and foods for something that many have termed the 'ancient future' (Mushita and Thompson 4). They are reintroducing traditional medicines such as creosote and wild foods into their contemporary diets to regulate blood sugar and reduce the effects of diabetes; more recently, through a series of successful lawsuits, they have been regaining their water rights and increasing the outputs of their agriculture as they return to traditional foods, medicines and seeds (Adamson 'Ancient Futures'). In the midst of Phoenix, Arizona where many continue to live today, they are illustrating how to gather wisdom from the past to face and meet today's most pressing social and environmental challenges.

King Clone: visualising time, constellating new humanities practices

Guided by a set of 'Common Threads' found on the international HfE website, HfE researchers are creating innovative projects to address problems that are 'shaped equally by the past, present, and the future'.[8] Like photographer Rachel Sussman's book, *The Oldest Living Things in the World*, an arts-science study of species that have been alive for 2000 years or more, Life Overlooked was designed to guide students towards a deeper, more passionate understanding of all species, charismatic or overlooked, and the ways in which our planetmates support a long-lived global ecosystem. To facilitate this goal, I adapted Environmental Creative Nonfiction for students at Arizona State University's urban campuses. Although most students had grown up in Phoenix, many did not know the names of a single cactus or the most common plants, including creosote. At ASU's Polytechnic campus, a model of sustainable architecture and drought-tolerant landscaping, I take my students out onto the campus for walks to introduce them to creosote, mesquite and prickly pear. Creosote, I tell them, having learned this from Rachel Sussman, can live up to two years with no water, and regularly survives days when the temperature is over 115 degrees Fahrenheit; the plant has also been able to survive extreme human activities, including nuclear bomb testing conducted in the Nevada desert (Sussman, *The Oldest* 18–22). This is why creosote is a plant well suited for desert landscaping on a campus seeking to be a model of urban sustainability. My class discussions of creosote, hopefully, are an illustration of how studying one overlooked plant can make evolutionary processes more visible.

Sussman has made creosote visible by photographing the oldest known individual, affectionately anthropomorphised because of its great age, as 'King Clone'. This large individual plant grows near Death Valley in the California stretch of the Mohave Desert. Sussman chooses to photograph creosote and other long-lived species because they are something of a palimpsest, or layered story that tells of hundreds of years of history and evolution. Her images reveal how all of the stories she tells (about the species she photographs) are 'interconnected – and in turn, inextricably connected to us all' (Sussman, *The Oldest* xii).

While I was teaching O'odham high school students in the 1990s, I learned that stories about creosote in the oral tradition do something very like Sussman's anthropomorphised photos of 'King Clone'. Creosote is central to O'odham cosmological understandings of how the world came to be. O'odham elders, interviewed by Gary Paul Nabhan for *Gathering the Desert*, say that once there was nothing on the land, and the world was dark. Then, Earth Maker's spirit began filling the planet. He took soil, and flattened it in his hands. From that soil, the first green thing, creosote, grew. From the plant's tiny, resinous leaves, the first animal, a tiny insect, emerged. This insect used the plant's resin to protect its own body and then to create the mountains (Nabhan 11). The story is remarkable for its botanical accuracy and for its depictions of the entangled relationships between insects, plants, humans and their surrounding ecosystems. Creosote is among the first flowering plants to evolve on the planet. Some pieces of dead creosote have been radiocarbon dated back 17,000 years (Sussman, *The Oldest* 22). Thus, O'odham stories about creosote are far more than interesting or entertaining cultural artifacts. They offer insight into human–non-human relationship over deep time, or more specifically, to the relationships between the Hohokam and their descendants' use of creosote as a kind of 'drugstore' (Nabhan 14–15). Like the name 'King Clone,' the stories 'anthropomorphise' creosote, making it a 'creator being'. As a creator being, the plant calls attention to its long-lived evolution and adaption to extreme conditions. As a creator being, creosote is a seeing instrument because it helps make the connections between biogeochemical processes and the 'well-being' of all species on the planet more visible (Adamson 'Humanities' 136).

The natural history of creosote also points to deep time, scale and common ancestors. Scientists have long known, as Dorion Sagan observes, about the 'massive evidence – from fossils, biology, morphology, organismal behaviour, biogeochemistry, ecology, and cosmology' – that makes it clear that 'life has been evolving since its inception over 3.5 billion years ago' (Sagan 113). Algae is at least 500 million years old and plants began developing after green alga started making its way on to land and becoming more complex. From this early life, mosses began to develop, and 100 million year later, ferns and horsetails. Fifty million years later, plants with seeds, cones, conifer and spruces; then, 140 million years ago, flowering plants appeared. In a study of more than 20,000 plant fossils, Daniele Silvestro and her team of scientists have explored why plants are figuring so importantly in recent studies of species adaption that aim to understand the effects of mass extinction events on plant diversity, biodiversity and Earth's ecosystems. In general, plants have been particularly good at surviving and recovering through these events. For example, the Cretaceous-Paleogene mass extinction, caused by the impact of an asteroid off the Mexican coast some 66 million years ago, dramatically reconfigured terrestrial habitats and led to the extinction of all dinosaurs except birds. Surprisingly, however, it had only limited effects on plant diversity. Flowering plants (angiosperms), in particular, did not suffer from increased extinction, and shortly after the impact they underwent a new rapid increase in their diversity. These evolutionary dynamics contributed to making flowering plants dominant above all other plant groups today (Silverstro et al. 1).

There are at least 350 thousand flowering species in total in the world today, writes Silvestro and her team, many the source of our foods and medicines, including potatoes, mints, lilies, orchids, grasses, cacti, roses and more. Botanists tell us that flowering plants changed the world, and made possible the diversity of life and other kinds of lifestyles for other organisms, including insects, fungi and humans who depend on the resources they derive from plants. Later, Darwin was among the first to suggest that plants might have something like a 'nervous system' and that they might have a common ancestor. Indeed, as botanist Rainier Stahlberg writes, Darwin was also one of the first to research the role of electric currents in plant movement. He was animated by some of the same questions first spurred by scientific discoveries that plants such as pumpkins with their trailing vines responded to stimuli. Darwin's research was considered controversial and was rejected by many. However, by 1907, some scientists were referring to plants as having nerves (Stahlberg 7), and today, Darwin's hypothesis about plants having a control centre for behaviour dispersed across their root tips, or something that scientists are terming a 'root-brain,' is coming to be widely accepted (Garzón and Keijzer).

However, plant 'root-brains' are not something that humans are typically used to seeing or thinking about, and thus, not something that most humans *can* see, unless through some kind of seeing instrument, sometimes a microscope, or sometimes a story. This suggests some of the reasons why a Life Overlooked project is needed – to teach students how to look for things they may not typically see because they are not looking. This also suggests some of the reasons environmental humanists are saying that we need new 'constellations of practice' to help us see deep time, scale and intricately entangled relationships between human and nonhumans that may be relatively invisible to most humans.

One of these new practices is time-lapsed photography. For example, the PBS programme *Nature*, in an episode titled 'What Plants Talk About,' shows that plants are not sedentary and that they can actively respond to nutrients and predators, and nurture their young. Scientists use time-lapsed film and photography to speed up or slow down time in ways that allow humans to *see* grass growing, roots foraging for food and fungal networks feeding their young ('What Plants Talk About'). Another example is Louis Schwartzberg's National Geographic 3D IMAX film, *Mysteries of the Unseen World*. Schwartzberg uses high-end time-lapsed cinematography to make time and plant processes visible. He shows his audience how organisms emerge from the soil and vines survive by creeping from the forest floor to look at the sunlight. Schwartzberg's aesthetics and film techniques nurture his audience's curiosity and wonder. These tools can be used, together with traditional humanities tools, storytelling and narrative, to constellate multimodal practices that might be used to scale up the numbers of teachers and students learning more about how anthropogenic and biogeochemical processes are always in relationship and can be made more visible.

As the lead developer of the HfE international website team, I have chosen the word 'constellation' to discuss how humanists are working with both traditional and new methodologies and practices not only because of my own longtime scholarly interest in ancient observatories, almanacs and oral narratives as seeing instruments, but because it is associated with astronomy, constellations and the stars to which most human cultures have linked their ancient or founding cosmologies (Adamson 'Humanities' 136–138). When I took charge of the HfE website in the summer of 2014, and began creating a digital platform that would work for the Life Overlooked project, I was guided by my past experiences teaching in digital environments. I called upon my previous experiences teaching environmental creative nonfiction courses with both the traditional writer's journal and with new technologies. In the early 2000s, on my small rural campus, I may have been among the first university professors to work with electronic tools, one of which we called the 'webboard'. I offered my courses in different semesters either face to face (f2f) or in hybrid formats. After I arrived at Arizona State University, I taught in f2f, hybrid and fully online formats. New technologies allowed my students to share and workshop their creative nonfiction essays both inside and outside of the classroom.

For the Life Overlooked project, working with HfE web designer, Patricia (Patty) Ferrante, we created two different template Wordpress formats for students who would be creating web pages for the courses they were taking with LeMenager, Sandiland or me. One template was for the student new to Wordpress and easier to use than the other template for the more digitally advanced. These two templates offered students a range of digital tools from the very simple (uploading images and text, tagging, geolocation, etc.) to the more complex (buttons, tiles, embedding videos, etc.).

Life Overlooked and citizen humanities

Anthropologist Eduardo Vivieros de Castro has elaborated on Indigenous philosophies and stories about anthropomorphised plants and animals considered 'persons' with stories that become an imaginative force for thinking about how the Earth was formed through a series of relations, often predatory and violent, through geologic and biospheric time and how these insights might assist humans in adapting to a rapidly changing world (Adamson 'Source'; Castro 469-488). Donna Haraway has called stories about Mother Earth, Gaia, Grandmother Spider and Indigenous stories about anthropomorphic or sentient persons, 'collected things'. Haraway's term gives presence to an entangled myriad of temporalities and spatialities, or 'entities-in-assemblages – including more-than-human, other-than-human, inhuman' (Haraway 160). With the O'odham philosophies and stories about Earth Maker, and Sussman's work on King Clone in mind, my students and I looked around the urban laboratory of our campus to discover how overlooked species could be understood as an entangled myriad of assemblages, or collected things that might be seen if we knew how to look.

Since we would be attempting to represent human relationships to common species like crickets, grass, starlings and pigeons in a digital environment, I asked my students to explore websites focusing on citizen science and community activism on behalf of overlooked species. They studied sites such as Maya Lin's 'What is Missing?', Anna Tsing's 'Matsutake Worlds' and a website they discovered on their own, The Biodiversity Group's 'Life Overlooked'.[9] They learned the Biodiversity Group had long been entangled in important ways with the concepts we were piloting. Completely independently from our project, The Biodiversity Group, a collective of researchers and activists, had been using the phrase 'Life Overlooked' on their Facebook page since 2012. This group networks citizens, scientists and photographers interested in 'citizen science' on behalf of the world's snakes, lizards and amphibians and their website became an important model for our own work.

After asking students to research websites, I asked them to read Henry David Thoreau's long essay, 'Walking,' for its insights into creativity, movement, migration and mobility. We read Álvar Núñez Cabeza de Vaca's *Castaways* and Gary Nabhan's *Gathering the Desert*. These texts allowed me to encourage students to explore their relationships with human and nonhuman others through observant walking and sitting on campus, and deep thinking about the ways language, and even single words, shape what we see, and don't see, and how we experience a nonfamiliar species (like Cabeza de Vaca's 'prickly apple'). We also read nature writer Lyanda Lynn Haupt's *The Urban Bestiary* and journalist Elizabeth Kolbert's *The Sixth Extinction* to learn more about scientific discourses surrounding accelerating environmental change and extinction rates. Mid-way through the course, I asked students to begin organising all the images and information or other elements they would use to create a web page of their own.[10]

But first, with citizen science in mind, we needed to think about how we might become 'citizen humanists'. We read Noel Castree's 'The Anthropocene and the Environmental Humanities: Extending the Conversation,' which helped us to define citizen humanities as something that might take a form like Castree's definition of 'engaged analysis'. This kind of analysis, Castree argues, would bring the best practices of the humanities and social sciences to bear on the most important conversations about global and local environmental change both inside and outside the humanities (Castree 244). My goal with assignments was to invite students to ask questions such as 'Why have "bestiaries – compendiums of animal lore and knowledge", common in medieval times in Europe, been all but lost?' (Haupt 5). Other questions we developed in class included, 'Why have older understandings of the natural richness of biodiversity, that can be found in oral and written forms in every culture, but especially in many Indigenous cultures, been largely dismissed or forgotten?' and 'How might "citizen humanists" bring these recorded cultural memories back into the scientific record to tell richer, fuller stories about species that are smaller than the human hand or commonly overlooked?' With my own research on the oral traditions, science and poetry associated with creosote as example, I asked students to innovate a 'citizen humanities' from constellations of traditional humanities practices including reading and writing fiction, poetry and nonfiction, and arts practices such as drawing, and from new media tools and arts such as photography, graphics or the production of short documentary films. Students were also required to delve into scientific research and to understand the ecological importance of their chosen species and its role in urban environments. Each student was then asked to write an engaging, creative nonfiction narrative about the species they chose and upload it, along with photos, art and other materials that would help make their overlooked species as well known as a polar bear or a whale to our digital platform. Students learned fascinating new information from reading each other's web pages. For example, they learned that some people may be eating crickets as protein in the next century to lower the carbon footprint of animal feedlots now providing the world with meat and that pigeon poop can literally dissolve cathedrals.[11]

In completing their web pages, students tested how humanists can intervene digitally and narratively in the way the general public approaches an understanding of cactus wrens and prickly pear indigenous to the Sonoran Desert that can commonly be found in urban environments, or non-indigenous species that have migrated, accidently or purposefully, to cities such as Phoenix. These include geckos, grackles and starlings. One student, Patrick Dennis, not only learned the lore and science of the prickly pear, as a collected thing, a food and a plant, he learned that the plant indigenous to the Sonoran Desert had travelled around the world to become a cherished 'local food' in places as far away as the Penghu archipelago in Taiwan. In 'Prickly Pear: Enemy and Ally,' he writes that a parasitic insect that lives on the prickly pear, the cochineal, literally filled the coffers of monarchies during colonisation after it was discovered that the insect's crushed body contained a red colour so bright that kings declared that only royalty could wear clothing made with the dye.[12] Dennis sought to understand how stories about prickly pear and its parasite, the cochineal, might be employed to broaden the application of principles of affective attachment, social justice and environmental sustainability, to make one non-charismatic species, and all the other species in relationship with it, including the human, more visible. Like Anna Tsing's Matsutake Worlds website, my student's web pages uncovered whole economic and botanical systems that go largely unseen in this era that is increasingly referred to as the Anthropocene.

In a new epoch, can humans learn from past civilisations or 'from past challenges of resource scarcity' to build ancient futures? How quickly and 'how successfully' can they respond to accelerating environmental and widening social disparities? (Holm et al. 984). These were some of the questions that Life Overlooked sought to bring into view by using common species as seeing instruments. By constellating ancient, traditional and new humanities practices each teacher and student involved in the project sought to pilot a citizen humanities that might complement what we mean by citizen science and contribute to the most important conversations about how to respond to global and local environmental change. Each aimed to create more awareness of our connections to the planetmates that support us within the long-lived global ecosystem we call Earth and to provoke deeper thinking about the kinds of futures we will build to make cities livable.

Notes

1 Pueblo Grande is one example of many pyramidal-sub-structured mounds that can be found throughout Central and North America, from the Maya ruins to the Mississippian culture mounds at Cohokia discussed by Linda Hogan in 'Backbone: Holding Up Our Future' (Chapter 2).
2 See Pueblo Grande Museum and Archeological Park, www.phoenix.gov/parks/arts-culture-history/pueblo-grande/. Pueblo Grande Observatory ruins are listed in the US National Register of Historic Places reference #66000184.
3 See North American Observatory, http://hfe-observatories.org/observatories/north-american-observatory/
4 The common project goals of Adamson, LeMenager and Sandilands are explored in the co-authored essay, 'Citizen Humanities: Teaching "Life Overlooked" as Interdisciplinary Ecology' published in *Resilience: A Journal of the Environmental Humanities* (forthcoming). See the Life Overlooked project, and each of the three template syllabi created by LeMenager, Adamson and Sandilands, here: http://hfe-observatories.org/project/life-overlooked/.
5 See the international Humanities for the Environment (H*f*E) website, http://hfe-observatories.org/. Designed as a 'laboratory' or 'research' and 'networking' space to forge 'new constellations of practice', the site brings traditional humanities tools together with digital tools, curation, mapping, short videos, visualisation, etc., to envision how humans might create a livable 'future we want'. See Adamson, Chapter 1.
6 For more on 'walking into knowledge', one of the themes of this collection, see Michael Davis, Chapters 4 and 12.
7 See 'Cabeza de Vaca Trails in Florida, New Mexico and Arizona', www.floridahistory.com/vaca-3.html
8 See H*f*E Common Threads: http://hfe-observatories.org/common-threads/
9 See 'What is Missing?' www.whatismissing.net, Matsutake Worlds www.matsutakeworlds.org and The Biodiversity Group http://biodiversitygroup.org
10 See Life Overlooked: http://hfe-observatories.org/project/life-overlooked/. Joni Adamson's syllabus can be found here: http://hfe-observatories.org/wp-content/uploads/2015/12/Adamson_Life_Overlooked_Syllabus.pdf
11 See Emily Early, 'The Cricket Chronicles', http://hfe-observatories.org/project/crickets/ and Erin Barton, 'Columbidae: The Human and Natural History of the Pidgeon' http://hfe-observatories.org/project/pigeons/
12 See Patrick Dennis, http://hfe-observatories.org/project/prickly-pear-destroyer-of-worlds/

Works cited

Adamson, Joni. *American Indian Literature, Environmental Justice, and Ecocriticism: The Middle Place*. Tucson: University of Arizona Press, 2001. Print.

——'The Ancient Future: Diasporic Residency and Food-based Knowledges in the Work of American Indigenous and Pacific Austronesian Writers', Special Issue: 'Migrants and their Memories'. Guest Co-Eds K.T. Tee, Ayeling Wang and I-Chun Wang. *Canadian Review of Comparative Literature*. 42.1 (March 2015). 5–17. Print.

——'Humanities', *Keywords for Environmental Studies*. Eds Joni Adamson, William A. Gleason and David N. Pellow. New York: New York University Press, 2016. 135–139. Print.

——'Source of Life: *Avatar*, Amazonia, and an Ecology of Selves', *Material Ecocriticism*. Eds Serenella Iovino and Serpil Opperman. Bloomington, IN: University of Indiana Press, 2014. 253–268. Print.

Adamson, Joni, Stephanie LeMenager and Catriona Sandilands, 'Citizen Humanities: Teaching "Life Overlooked" as Interdisciplinary Ecology', 'Green Humanities Lab.' Special Issue. *Resilience: A Journal of the Environmental Humanities* (forthcoming 2017). Print.

Anderson, Mark T. and Lloyd H. Woosley Jr. 'Water Availability for the Western United States – Key Scientific Challenges', U.S. Geological Survey, U.S. Department of the Interior, Circular 1261, 2007. Print.

Castree, Noel. 'The Anthropocene and the Environmental Humanities: Extending the Conversation', *Environmental Humanities* 5 (2014): 233–260. Print.

Castro, Eduardo Batalha Viveiros de. 'Cosmological Deixis and Amerindian Perspectivism', *Journal of the Royal Anthropological Institute* 4.3 (1998): 469–488. Print.

Cabeza de Vaca, Álvar Núñez. (1555). *Castaways*. Ed. Enrique Pupo-Walker, trans. Frances M. López-Morillas. Berkeley, CA: University of California Press, 1993. Print.

Garzón, Paco Calvo and Fred Keijzer. 'Adaptive Behavior, Root-brains, and Minimal Cognition', *Adaptive Behavior* 19.3 (June 2011): 155–171. Print.

Haraway, Donna. 'Anthropocene, Capitalocene, Plantationocene, Chthulucene: Making Kin', *Environmental Humanities* 6 (2015): 159–165. Print.

Haupt, Lyanda Lynn. *The Urban Bestiary: Encountering the Everyday Wild*. New York: Little, Brown and Company, 2013. Print.

Holm, Poul, Joni Adamson, Hsinya Huang, et al. 'Humanities for the Environment – A Manifesto for Research and Action', *Humanities* 4.4 (2015): 977–992. Web.

Kolbert, Elizabeth. *The Sixth Extinction: An Unnatural History*. New York: Henry Holt and Co., 2014. Print.

Martin, Patricia Preciado. *Songs My Mother Sang to Me*. Tucson, AZ: University of Arizona Press, 1996. Print.

Mushita, Andrew and Carol B. Thompson. *Biopiracy of Biodiversity: Global Exchange as Enclosure*. Trenton NJ and Asmara, Eritrea: Africa World Press, 2007. Print.

Nabhan, Gary. *Gathering the Desert*. Tucson: University of Arizona Press, 1985. Print.

Ross, Andrew. *Bird on Fire: Lessons From the World's Least Sustainable City*. London: Oxford, 2011. Print.

Sagan, Dorion. 'Evolution', *Keywords for Environmental Studies*. Eds Joni Adamson, William Gleason and David N. Pellow. New York: New York University Press, 2016. 116–118. Print.

Silvestro D., B. Cascales-Miñana, C. D. Bacon and A. Antonelli. 'Revisiting the Origin and Diversification of Vascular Plants through a Comprehensive Bayesian Analysis of the Fossil Record', *New Phytologist*, 2015. Web.

Stahlberg, Rainer. *Plant Signaling and Behavior* 1:1 (January/February 2006): 6–8. Print.

Sussman, Rachel. *The Oldest Living Things in the World*. Chicago: University of Chicago Press, 2014. Print.

——'The World's Oldest Living Things', TED. July 2010. Web.

Swartzberg, Louis. 'Hidden Mysteries of the Natural World', Ted Talks. March 2014. Web.

'What Plants Talk About.' *Nature*. PBS. 3 April 2013. Web.

Zepeda, Ofelia. *Ocean Power*. Tucson: University of Arizona Press, 1995. Print.

9 Environmental rephotography

Visually mapping time, change and experience

Mark Klett and Tyrone Martinsson

Rephotographs are photographs that intentionally repeat earlier images, usually from the same vantage points in space, in order to allow a detailed comparison between views. Rephotography forms a dialogue with time, history and memory where multiple photographs made at different times become points of comparison that anchor historical documents to physical locations. As a fieldwork method rephotography literarily takes visual artifacts back to where they were once produced and offers a unique opportunity to see the relationship of historical records to the spaces they represent. More recently as attention turns to the effects of global warming and climate change in what is largely described as the Anthropocene era, rephotography can serve as a tool for comparative studies showing clear evidence of human-caused change. In combination with other data the results may be used to help humanists, scientists, policy makers and the general public visualise the effects of such changes. One result of rephotography is to influence human behaviour by challenging the expectations of place that a limited viewpoint in time provides. This is possible because rephotographs connect pictures from one time to another, forming a new entity extending beyond past and present to pose questions about the future.

This chapter will explore the process of rephotography by looking at how it was used in several studies. First, three projects identify physical and cultural changes in the American West. We will describe briefly the methods involved, with the intent of illustrating what can be learned from the photographs. Next we turn to a project of environmental concern from the opposite side of the globe and in a very different environment. This project from the Arctic illustrates in dramatic fashion how rephotographs are used to record receding glaciers and the effects of climate change. The work involves a merger of creative practice, research and scientific interpretation. Taken together these projects present examples of how rephotography can be used to examine the environment and human interventions, providing perspectives on both time and change.

Environmental rephotography

Environmental rephotography is the practice of using rephotographs as a tool for addressing environmental issues and their perception. The work combines the humanities and the arts in a collective action needed for tackling the challenges of visualising the future. Environmental rephotography as practised by the authors of this chapter communicates the seriousness of environmental matters but balances doomsday perspectives by emphasising that hope exists through awareness and actions yet to be taken. James Hansen, in his book, *The Storms of My Grandchildren: The Truth about the Coming Climate Catastrophe and Our Last Chance to Save Humanity*, poses an important question about how to convey the consequences of continued exploitation of natural resources. Hansen, one of the world's leading climatologists, writes:

> What will the world be like if we go down this route? The science tells us exactly what we could expect to happen on Earth if we continue our business-as-usual exploitation of fossil fuels. I've referred to it earlier: the Venus Syndrome. But how to portray the horror of that devastation in a way beyond graphs and numbers and phrases we have heard before, like 'climate disaster'? (260)

Environmental rephotography provides one solution. It is a documentary medium using a visual language that is shared by humanists, artists and scientists alike. Photographs make data visible and connect observations to real places. Rephotographs may convey stories about these places and the serious consequences of human impact on earth. The narrative capabilities of the medium often employ language common to art practice, such as the use of poetic structure to express a human relationship to the land. But this practice also has an early history in environmental literature. Examples include the writings of Henry David Thoreau and John Muir. In particular John Muir expressed the importance of making connections about nature that linked the local to global (248). Aldo Leopold expressed the need for an aesthetic component to wilderness policy (Cardenas n. pag.). Decades later Rachel Carson was able to bring the issue of pesticide abuse into the public sphere by helping readers experience the painful loss of a world without birds. Scientists had known about the problem for years. But it took a writer who could link cause and effect in a way that a layperson could understand before public awareness created meaningful changes in behaviour.

More recently writer Barry Lopez laid open the poetic concept of *beauty* as a path of resistance when praising Subhankar Banerjee's 2003 book, *Arctic National Wildlife Refuge, Seasons of Life and Land*. What is being resisted in this instance is the assumption that continued exploitation of natural resources is necessary at all costs – or more importantly that it is inevitable. Lopez, in a short endorsement paragraph appraising Banerjee's book, turned to the complexity of beauty in contemporary culture, claiming it as 'our era's most suppressed topic of conversation' (back cover). This brief remark is at the heart of environmental photography. How do aesthetics and notions of beauty enter the discussion of a changing environment?

As the Humanities for the Environment (HfE) project on which this book focuses, and many other projects in the environmental humanities are demonstrating, the arts and humanities are in a position to critique the results of both human actions and assumptions.[1] They command an argument for not only *looking at*, but also *seeing* and *feeling* the surrounding world in all its complexity. They make visible the beauty of nature, the effects of loss and change, and connect them to the effects of human actions in a way that pure data itself cannot. This position can be in contrast to a disengaged perspective that reduces change, however horrific, to a quantitative summary.

Historian William Cronon reminds us that wilderness itself is a construct of culture. What art can do, and rephotography in particular, is to help visualise the numerous ways, for better or worse, the human species has constructed an environment in its own image. To restore aesthetics as a topic of environmental discussion it is worth exploring strategies that make possible a broader perspective on where we have been, where we are at the present moment and where we are headed. Rephotography is a perfect medium for such a reckoning.

The American West: a unique place to test rephotography's methods and results

The earliest photographs from the American West focused on the region's arid landscapes. The geological surveys of the 1860s operated under government sponsorship and employed teams that included scientists, soldiers, mapmakers and photographers in an empirical process of field observation and reporting. The photographs were disseminated in reports created for eastern audiences unfamiliar with the region's dryness, yet curious about the potential of its open lands. The survey photographers often focused on water and its sources. The power of flowing water in a dry environment had been the primary force in sculpting the geography of the West. Water as a resource for agriculture and urban growth quickly became a primary concern for an expanding national enterprise. As a growing population settled the West, water's availability, or lack thereof, shaped the political landscape as well as its physical appearance. Thus the survey photographs of the mid nineteenth century represent a dual record that continues to be relevant to the present day: they record the shape of the land as it once was, and they reflect the evolving cultural expectations that framed the land as a material resource and shaped its future.

Today these early photographs can be used to compare changes to the land and examine the influence of cultural interventions that have affected change. This was the premise behind two projects that used early survey photography as a basis for making new photographs. The Rephotographic Survey Project and the Third View Project compared the West of the nineteenth century to the present. The Rephotographic Survey Project (RSP), 1977–1980 (Klett et al. *Second View*), relocated the western survey photographs originally made for the King, Wheeler, Hayden and Powell expeditions of the 1860s and 1870s. These surveys predated the advent of the U.S. Geological Survey and were the first federally funded groups to explore and map the intermountain west including the current states of Arizona, New Mexico, Colorado, Utah, Nevada and Wyoming. The surveys employed scientists, topographers, anthropologists and photographers among others. The photographs made for the surveys were the first of any kind in these regions and as such form a kind of visual baseline for the way the landscape looked before modern intervention. The RSP attempted to reoccupy the original photographs' vantage points as precisely as possible, and also the lighting created by the time of day and time of year the originals were made. The techniques, limitations and observations about the process of making rephotographs for this project has been written about elsewhere and has been the basis for many projects that followed.[2] For the purposes of this chapter there are several environmental examples worth mentioning. First, the largest scale, human induced changes to the West involves water impounding and distribution. Examples include the construction of large reservoirs such as Green River as seen from the Vermillion Cliffs (Figure 9.1).

Photographs from Echo Canyon Utah (Figure 9.2) show transportation to be an important factor in change. Here a river was displaced by construction of an interstate highway and a natural rock formation destroyed by railroad building. Factors that made growth of the West possible such as water distribution and transportation infrastructure are rarely seen as directly as in these photographs.

Figure 9.1 Left: Green River Cañon, Upper Cañon, Great Bend; right: Flaming Gorge Reservoir from above the site of the Great Bend.

Figure 9.2 Left: Hanging Rock, foot of Echo Canyon, Utah; right: Hanging Rock, foot of Echo Canyon.

One may expect environmental change in photographs made over a century apart to appear one directional, and to show a constant, degrading, effect. But in some cases, rephotographs defy expectations. A photograph of Georgetown Colorado (Figure 9.3) shows that almost all the trees surrounding the mountain town had been clear-cut in 1875.

In the twentieth century this is reversed and the tree stands have grown back to a more natural state. Of course the photographs do not explain the contemporary energy sources that have replaced woodcutting, but they hint at that by showing the interstate highway that cuts across a distant hillside.

Overall, the RSP repeated 122 nineteenth century survey photographs and the results show patterns of water use, transportation, urbanisation and cultural intervention. The pictures also show vegetation changes and erosional changes that are the result of natural and human-caused forces. The image pairs are effective at showing differences in this environment over time, but one limit of the project is that it depends on what was captured by the original photograph. When the survey photographers made their photographs they unwittingly laid down a template for revisiting the subjects again and again over time. But they could not have known the changes in store for the lands they were documenting, nor where these changes would occur.

The Third View Project, 1997–2000 (Klett et al. *Third Views*), began on the twentieth anniversary of the RSP and was designed to address some of the first project's limitations. Third View explored technological advances that made it possible to collect new data in the field and attach this information to the repeated photographs. This opened a door for a wider range of interests combining art and humanities methods. After the rephotographs were made at each site, new photographs of the landscape were taken outside of the original view; video footage was recorded around the site or from the journey to the location. Sound recordings and oral history interviews were recorded with people connected to sites. Other made-for-computer images such as panoramas and animations were created to place the scenes in a contemporary setting and show what is behind the camera or outside of the picture frame. In addition, artifacts (contemporary and not of archaeological value) were collected at significant locations. Thus,

Figure 9.3 Georgetown, Colorado.

besides physical change, the project shifted its focus to include the context surrounding each view, the evolving perceptions of place, shifting mythologies of the West and the many issues of documentary practice.

Fieldwork collected a large array of visual and aural documents, and later these components were edited and combined into interactive presentations. The project launched a website (www.thirdview.org) containing many of the less memory-intensive works. The fully edited materials were presented in an interactive DVD that was published and included in the book jacket. In the DVD, viewers may access the video, sound files and other images that connect places to events, people and ideas. In addition, there were several significant changes to the techniques used to view rephotographs themselves.[3] Third View represented a shift in the working methods of rephotography to a more inclusive and interactive process that provides a broader context for how rephotographs visualise time, and subsequently how they can be interpreted. This approach enabled the work to tell stories about local changes and narratives related to sites. For example at Green River, Wyoming, Third View found former landscape icons of the nineteenth century converted to common backgrounds by the town's urban development. Teapot Rock, once chosen by several nineteenth century photographers as a dominant feature, is now situated next to an interstate highway (Figures 9.4 and 9.5). This former icon of the West was being used as a local beer drinking hangout.

At Pyramid Lake in Nevada, O'Sullivan photographed tufa formations above this natural lake fed by the Truckee River. In the rephotograph from 1979 the water level had dropped over twenty metres due to water diverted to feed irrigation projects south of Reno. The lower levels and adverse environmental consequences prompted litigation from the Paiute Indian tribe, the lake's owners. Following the success of those actions a higher lake level can be seen in the year 2000 third view (Figure 9.6).

In urban areas such as Salt Lake City, the first settlement was little more than tents on the horizon when O'Sullivan first photographed there in 1868. In 1998 the development reached the edge of the original photograph and an audio interview captures a resident explaining that there's little land left to purchase in Salt Lake, and this new development, on a hillside, was one of the last, best options (Figure 9.7).

Figure 9.4 Left: Tertiary Bluffs near Green River City, Wyoming; middle and right: Teapot Rock and the Sugar Bowl, Green River.

Figure 9.5 Still image of artifact found at Teapot Rock, Green River Wyoming.

Figure 9.6 Left: rock formations, Pyramid Lake, Nevada; middle: Pyramid Isle, Pyramid Lake, NV; right: Pyramid Island from the tufa knobs, Pyramid Lake, NV.

Figure 9.7 Left: Wasatch Mts, Salt Lake City, Utah, Camp Douglas and east end of Salt Lake City; middle: Salt Lake City; right: new housing development, 'Dorchester Pointe', Salt Lake City.

At Logan Springs, Nevada, we see the opposite of urban growth. An abandoned house depicts pioneers assembled at a former mining venture in Timothy O'Sullivan's 1872 photograph (Figure 9.8). The rephotograph shows a contemporary dwelling made from the stones of the original structures. The DVD's video takes a tour of this house and raises unanswered questions about the people who lived there and why they left. These parallel questions about the

Figure 9.8 Left: water rhyolites near Logan Springs, Nevada; right: water rhyolites near Hiko, NV.

original occupants of Logan Springs and why they left. The effect is to link the past to the present in a mystery of disappearance, one that leaves doubt about the permanency of present day interventions.

Rephotography of a drowned landscape and the adaptation of methods

Both RSP and Third View tried to relocate an original photograph's vantage point as precisely as possible. This was made possible because dry western lands change slowly and photographic subjects can usually be found. Precise relocation has become a convention of accurate rephotography methods, and it has two functions. First it enables more detailed comparisons between images. The camera acts as an optical instrument like a surveyor's transit, capable of measuring detailed physical change (Malde). Second, when accurate placement of the camera is paired with attention to recreating the lighting of the original photograph the pair functions as a more convincing duplicate.

But not all photographs are candidates for this kind of rephotography. What if the original vantage points no longer exist or cannot be found, let alone repeated with any degree of accuracy? How can we use photographs to visualise change over time? This is the question asked when reconsidering the Glen Canyon photographs of Eliot Porter, published in his 1963 Sierra Club book *The Place No One Knew*.[4] The waters of Lake Powell now submerge almost all of the places Porter photographed in Glen Canyon, and that was the point of the publication. Porter made his photographs knowing full well the lake's waters would soon drown his vision of Glen Canyon, and the Sierra Club wanted to call attention to what it considered an environmental tragedy.

In 1963, the Glen Canyon Dam was finished creating the second largest reservoir in the United States. The lake waters flooded over 150 kilometres of pristine but little known canyons of the Colorado River first surveyed by John Wesley Powell during his epic journey down the Grand Canyon nearly one hundred years earlier in 1869. The reservoir made urban growth in the southwest possible and is today a destination for motorised recreation. Yet, Glen Canyon

Dam's presence is still a flash point for organisations dedicated to the dam's removal, even 50 years after its construction.

The once-imminent destruction by water recorded by Porter also yielded a notable moment in the history of fine art photography. *The Place No One Knew* became a seminal publication marking another kind of confluence – of art photography and environmental activism. The book was an early example of the newly bifurcated and now established branch of 'Nature Photography,' photography that focuses on natural locations or animals in a wild state, without reference to human beings. The lake's waters further exemplified a moment of convergence for efforts to define wilderness and the politics of mass communication. The United States Congress passed the *Wilderness Act* a year after the dam's construction in 1964, stating that wilderness is 'an area where the earth and its community of life are untrammeled by man, where man himself is a visitor who does not remain' ('The Wilderness Act').

Porter's book was conceived of as a eulogy for a wilderness that was destined to be lost. His tools included his unique photographic vision and a cutting edge use of colour film, a rarity until then. Boldly, he eschewed the grand vistas of contemporaries like Ansel Adams, focusing instead on smaller, more contained scenes, closer to the camera. Porter's canyon landscapes rarely revealed the horizon or sky; they closed in on rock formations, the water in small pools or rivulets, sand, mud, animal tracks and plants. The saturated colours described a more intimate view of a hidden world contained within the red walls of Glen Canyon, a world revealed only by a close personal experience of place and expressed in a vision that evoked the powers of the picturesque. This was the experience that would be lost by the time the book landed in the hands of its readers. *The Place No One Knew* was intended to provoke public sympathy and outrage, and to prevent more such losses in the future. The Sierra Club considered the book a major part of its campaign to draw public awareness to the loss of wild places. The club fought against plans to build more large reservoirs, and Glen Canyon Dam became the last major project of the dam building era.

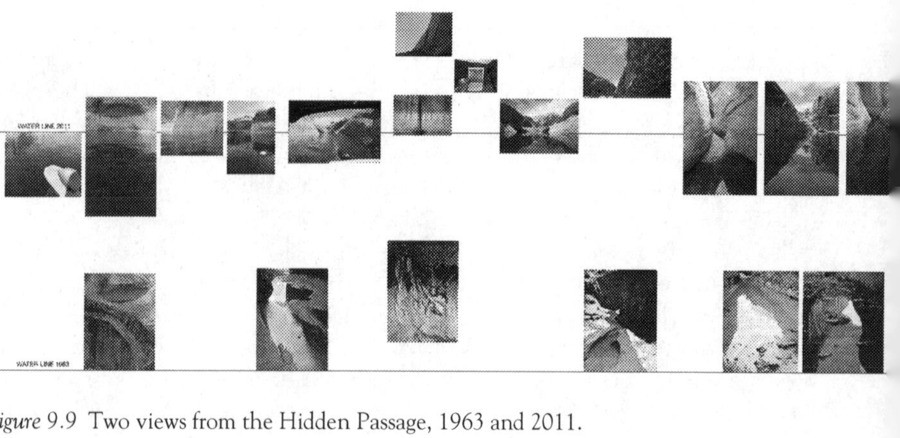

Figure 9.9 Two views from the Hidden Passage, 1963 and 2011.

Like RSP and Third View, Porter's historic images became points of departure for a project that revisits the places where Porter made his photographs. Unlike the two previous projects, however, new work is limited by its inability to locate Porter's vantage points. By necessity the effort involved altering the methods normally used to make rephotographs, and constructing entirely new images based on response to the subjects and the poetic content of Porter's original works rather than repeating his original vantage points.

Lake Powell can be considered a drowned river, one where contemporary visitors float on the surface of impounded water above a deep, hidden landscape. The lake's elevation has become a measure of time. Today, there is a white 'bathtub ring' that marks the rise and fall of water due to variations in yearly snowmelt and the lower levels brought by over a decade of drought. The reservoir is filling quickly with sediment. And a flood of visitors arrives with the seasons, travelling in armadas of houseboats, jet skis and powerboats. The project decided to reply to Porter's photographs and his Sierra Club book as artifacts of the past, treating the remnants left on the lake's shores by contemporary visitors as the corresponding artifacts of our post-dam era. Rather than measuring change, the project asks questions about how the place feels: how do we respond emotionally and aesthetically to a transformed landscape considered by many to have been violated? The goal is to contemplate lessons from Lake Powell as a way forward in the face of impending climate change in a drought plagued environment.

The new project has created several alternative strategies to address earlier photographs. When it's possible to relocate all or part of an original Porter photograph the project made pictures that referenced his vantage points. This can been seen in the photographs of the 'Hidden Passage' (Figure 9.9).

This series of extended views shows the water line of 1963 and the group of photographs Porter made while walking into a side canyon called 'Hidden Passage'. At the canyon's entrance, Porter pointed his camera upwards and captured a large lens-shaped alcove on the walls above. In 2011, we were able to find this same alcove half submerged by lake waters at the mouth of the same

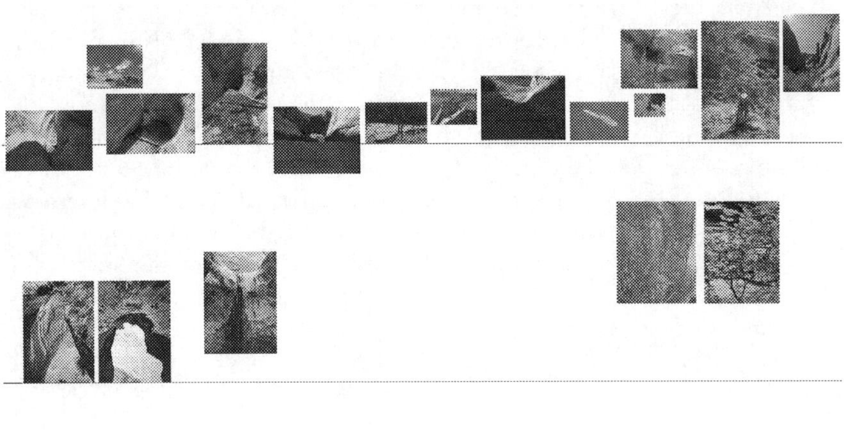

canyon as we approached by boat. Floating high above Porter's drowned vantage points we photographed the features seen in reference to the subjects Porter recorded below, 50 years earlier. We followed his reflections of rock formations in the water, though our views were often rimmed with lake debris. At the end of the Hidden Passage, the present day scene emerges to discover Redbud trees just as in Porter's photographs. The resulting group forms a linear sequence of photographs that traverses the same geography over two eras. The vertical distance represents the rising water level and also the separation in time. Porter's more picturesque photographs are countered by the realities of the contemporary lake's debris. Our intention was not to replicate the beauty of his work, but rather, to call attention to the loss of his vision and what was left in its place.

Porter's photographs were not meant to be literal documents of place. His vision was informed by the art practice of his day, and the language of photography combined both its descriptive capabilities with a metaphoric interpretation. For example, the reflections he photographed were not simply records of rock in water; they were signs of a natural beauty that was a redemptive force in the face of modern civilisation. Understanding Porter's original intent enables a more informed response in return. In one instance we paired similar reflections from completely different locations. Porter's tall rock formation is replaced by a figure in an alcove frequently visited by boaters, creating a contrast that points to the differences between time and place (Figure 9.10).

The examples cited above are not traditional rephotographs. But in the case of Lake Powell, some rephotographs can still be made. The waters are receding rapidly enough, and projected to continue with climate change induced drought, that large amounts of sediments are filling in side canyons. So, in addition to responding to Porter's photographs over the course of three years of fieldwork, we were able to make and repeat some of our own images as well. In 2012 we made a photograph to record the buildup of sediment at the north end of Lake Powell where the lake levels have receded dramatically (Figure 9.11).

In addition to lower water levels, the high sediment flows from a tributary stream have caused this far end of the lake to fill in. The earlier photograph shows this sedimentation as a flat brown plain covering what had been lake water during the years after the dam was built. This provided an excellent opportunity to visualise the lake's future. In a 2014 rephotograph made to match the earlier photograph, the lake can be seen to have receded even further and grasses have grown on the brown plain that had been barren only two years earlier. In this later view, the Dirty Devil River is seen carving a new channel. The forces that cut Glen Canyon are again at work, this time cutting through the plain that was deposited as a result of water dropping its sediment load. Natural processes are dominant once more, in spite of the dam, and will become even more evident as the water level lowers.

Figure 9.10 Left: Eliot Porter 1963; right: Mark Klett and Byron Wolfe, small alcove in side canyon, Lake Powell.

Figure 9.11 Left: the 'dead pool', where the Dirty Devil River dropped its sediment load in what had been the northernmost waters of Lake Powell; right: sediment plain covered by vegetation with new channel being cut by the Dirty Devil River.

At the opposite end of the earth: a study in glacial loss

The Arctic is a place where the dominance of natural processes is also evident. Glaciers have a long natural history and sometimes a cultural history as well. They are among the most treasured, beautiful features of the frozen north and are also good indicators of the effects of global warming for which they provide direct evidence. The following examples, illustrating rephotography, are from Arctic Svalbard and North West Spitsbergen National Park, a group of islands in the polar sea 800 kilometres north of Norway. This is one of the most historic areas in Svalbard and represents a unique mix of wilderness and cultural landscapes. Rugged and wild coastlines are filled with cultural remains framed by ice and mountains. This remote corner of the Arctic can be seen as a micro example of historical development in the modern world, especially Europe. Over the course of 400 years, North West Spitsbergen has been altered from a wilderness to whaling and hunting grounds, to a national park. Europe's first oil exploitation exterminated the Greenland Bowhead Whale, and later the burning of fossil fuels and subsequent chemical pollution further changed the ecosystem of Spitsbergen.

Its recent designation as a national park offers protection and new hope for a rewilding process, and a halt to the effects of human-caused changes. However, due to melting ice in the area, Svalbard will inevitably become a different land than the one first described by seventeenth century travellers.

The land discovered by the Dutch in 1596 was named Spitsbergen for the jagged peaks and pointed hills along its coastline. Svalbard, the Norwegian name given to Spitsbergen in the 1920s, had no Indigenous population and within ten years after its discovery the great abundance of natural resources was understood and exploitation of the area began. Exploitation of these resources lasted for about 200 years before whales and other sources became scarce and finally extinct. In 1671 Friedrich Martens, a surgeon from Hamburg who accompanied a whaling ship, created the first in depth description of North West Spitsbergen including fauna, flora and comments on features of the land, mountains, snow, ice, glaciers and whaling as well as the life, death and work of the whalers. Martens published a book about his travels to the west coast of Spitsbergen in 1675, with his observations and some of the first images of views he encountered.

In the long timeline of geological change, human observations on the life and death of glaciers must be considered relatively short. Today we are witnessing an unprecedented change in the Arctic environment due to human causes. Climate change is normally an earth system process that works outside the range of human records. Global warming, observed through the life of glaciers, offers a unique opportunity to visualise the larger scale effects of human-caused climate change. We can literally see the arrival of the Anthropocene, the age of humans and the new era in the earth's geological time scale, replacing the relative stability of the past 10,000 years of the Holocene. Glaciers offer ice records that can be interpreted by science as ice records of past climates long before human observations were possible. The loss of ice in the Arctic will most likely change our perception of the region from the one we inherited from our ancestors. We will also lose these valuable ice records.

Swedish glaciologist Per Holmlund has labelled glaciers 'jewels in the landscape'.[5] They appear with sparkling brightness in hues ranging from white to blue and green, murky brown and black. The story of ice and glaciers in the Arctic is literally a story of light. We know today that the sea ice and the larger glacial ice caps at the Poles function as part of the earth's cooling system with its capacity to reflect much of the heat coming from the sun. The sea is darker and absorbs the heat, leading to increased warming. On land the ice in glaciers reflect the light and the heat of the sun. They are the signs of a healthy earth system, and without them the land also turns dark and warms. This is of course an argument for the importance of glaciers, but it also points out a need for understanding the important role that wild places play in larger earth processes. In Svalbard, the scars on the mountain slopes, or trimlines, indicate the future view of the coastline as the glaciers vanish and their light fades. Environmental rephotography tells the story of the disappearing glaciers in Spitsbergen (Figure 9.12).

By the end of the eighteenth century, sketched images of Spitsbergen started to appear more frequently and in 1773 the first view of a specific glacier was

Figure 9.12 North-east coast of Amsterdamøya.

recorded by the John Phipps expedition. According to Phipps' journal (70), the sketch of the glacier was made by the midshipman Philippe d'Auvergne at the site where they arranged their scientific instruments, such as the Pendulum for time keeping and corrections, Equatorial Instrument the advanced mount for the telescope and astronomical Quadrant for the height of the sun, on Amsterdamøya in Smerenburgfjorden. The location of the artist was even marked on a map of the area made by the expedition. Phipps indicated in his journal that the glacier being viewed was across the fjord from where they made scientific observations. In the field, looking at the historical data and facing the view described, it's clear that there are two glaciers: the Kennedy glacier and the Fram glacier. Neither d'Auvergne nor Phipps named the two glaciers and these names are given at a much later date. Philippe d'Auvergne's sketch is most likely a combination of the two glaciers. Details in d'Auvergne's drawing reveal a burial ground in the foreground and rugged and jagged mountain peaks in the backdrop. In the north front of the site of the Kennedy glacier is Likneset, an earlier burial ground for the dead from Smeerenburg, a Dutch seventeenth century whaling station, one of the largest graveyards in Spitsbergen. Further comparison with d'Auvergne's drawing reveal a similarity with the sharply pointed hills in the background of the Fram glacier. It is worth noting that these two glaciers are probably the first to appear as ice features on early maps of the region from 1651 and onwards. The Norwegian photographer Anders Beer Wilse was in the area 135 years after d'Auvergne and unintentionally photographed the view of Phipps' expedition. In 2012 Anders Beer Wilse's works were rephotographed as part of the North West Spitsbergen Glacier Record Project, an ongoing research project since 2011 led by Tyrone Martinsson creating a visual mapping of north-west Svalbard tracing its recorded visual history, telling its story and setting up a survey system for photography-based glacier monitoring (Figure 9.13).[6]

Photographic records of North West Spitsbergen began about 1861 when the young medical doctor Axel Goës made a panorama of the south coast of Magdalena Bay while on Otto Torell's Swedish, scientific expedition. Today the panorama exists only as a lithograph made after the photographs, but due to its remarkable detail it was rephotographed in 2012 and offers a point of origin for the photography-based observations in the area.[7] In 1818, on David Buchan's British expedition, Charles C. Palmer, an officer in the Royal British Navy, trained in making watercolour drawings for surveying land, made a similar record of the same coastline (Beechey).[8] These two early examples provide connecting reference points but also make clear the differences between works of the artist hand and eye versus the artist with the instrumental recording capacity of a camera (Figure 9.14).

Comparing visual sources like these over time raises the question of whether the glaciers changed between, say, Marten's visit in 1671 and Buchans' visit in 1818. By creating sequences of images and visual timelines, we can arrive at a plausible answer to that question. Extending this inquiry into the more recent images, and arranging them from 1861 to the present, a clear sequence of dramatic changes in visual appearance of glaciers is revealed. In addition, a dramatic increase in change from about 1960 to the present is observed.

Magdalena Bay is one of the most historical areas in Svalbard with a 400 year old timeline of recorded observations. The bay was first visited and briefly noted by Willem Barentsz, a Dutch explorer on the ship discovering Spitsbergen. Philip Broke made the first detailed survey of the bay in 1807 where the features

Figure 9.13 View east across Smeerenburgfjorden from Amsterdamøya.

Figure 9.14 View south from north side of Magdalenfjorden.

including glaciers are described and marked. In Magdalena Bay contemporary maps note a location called Gravneset (the Grave Peninsula). Gravneset got its name from being the site of an old English burial ground from the whaling era. This location is one of the best landing beaches in North West Spitsbergen. It has also been a longstanding location for photography, drawing and painting of landscapes around the bay. It is located below the slopes of where a landmark glacier was marked by Philip Broke on his 1807 chart as, 'the hanging glacier,' once lay. Today, looking east into the bay, one sees the Waggonway glacier, and to the north the Buchan glacier, and to the west a new bay where the former Gully glacier lay. From Gravneset seven glaciers were visible until around 1950. Today only two remain in sight. One is entirely gone. Four have retreated beyond sight.

In 2011 the Glacier Record Project made its first visit to Magdalena Bay and the site of Broke's 'the hanging glacier'. The glacier is visible on Axel Goës's panorama and Charles C. Palmer's drawing. British Lieutenant Herbert C. Chermside photographed the glacier as part of Benjamin Leigh Smith's 1873 expedition to Svalbard. Nils Strindberg, the photographer of Salomon August Andrée's polar expedition also visited the fjord in 1896 and photographed the glacier.[9] In images taken on tourist cruises between 1905 and 1938, the glacier is visible and then it slowly melts away only to disappear completely in the 1960s. Today there is nothing to indicate where the glacier once lay (Figure 9.15). With the 'hanging glacier' gone we see a number of glaciers about to follow.

The remarkable Gully glacier has been a tourist attraction since the late nineteenth century and is one of the most photographed glaciers in the area. It used to have an impressive front into the sea on the south coast of the bay. There is a clear trail leading from the landing beach on Gravneset to the glacier. Photographs show its decline in size and its loss of visual appearance from 1861 to the present (Figure 9.16).

In the bottom of the bay is the Waggonway glacier that has been famous for its beauty and visual power framed by sharp and dark mountains. It used to cover the bay connected with two other glaciers, the Miethe glacier and the Broke glacier coming from the south and north mountain hills, respectively. By the 1950s the glaciers forming the massive ice wall in the bottom of the bay had started to show clear traces of change and 50 years later the three glaciers had lost their connection. Today the Waggonway glacier is about to disappear from view from the landing beach and its former size can only be imagined from the trimlines on the mountain slopes (Figures 9.17–9.20). The story of the glaciers in Magdalena Bay apply to all glaciers in North West Spitsbergen.

While working in the field in 2011 and 2012 we tried to come up with a way to describe the process of glacier life using the visual materials collected from archives and the new photographs, texts and drawings we were discovering and producing. 'Visual appearance' became a working term for describing the timeline of glaciers. We tried to connect each glacier to as many historical sources as possible. Most sources were photographs, some were text sources, and some were made by hand such as drawings, watercolours or maps. We then cross-referenced images and created sequences. If multiple sources were lacking, some images were simply paired for direct comparisons of change over time.

This produced a kind of collage technique that enabled us to combine historical sources up to and including contemporary visual records. These visuals expand work in the sciences and are linked to other works such as earth systems trend charts. All together, these materials help to visualise larger concepts such as 'the great acceleration'. This term is often used as a kind of 'meta concept' in reference to evidence accumulating in the sciences that is showing accelerating change in a number of earth systems since the 1950s. The changes in appearance of the Spitsbergen glaciers offer visual verification of this acceleration. The rephotographs made in Svalbard firmly establish the potential of the photograph as both a research document and an artistic representation. Further, the aesthetic qualities of the visual work are acknowledged while delivering the message of the research community to a global audience.

Figure 9.15 Magdalenefjorden.

Figure 9.16 Magdalenefjorden.

Figure 9.17 Magdalenefjorden.

Figure 9.18 Views from Gravneset.

Figure 9.19 View across Trinityhamna towards *Miethebreen*.

Figure 9.20 View, approximate camera position: N 79° 39'57" E 11° 02'57".

Conclusion

Looking at changes to the land and environment half the globe apart, one is tempted to reverse the normal expectations about time and change. It's logical to think that change is the product of time passing. Change happens because time happens, an expected result often spread over a century. But rephotography shows us that change is actually the true measure of time. When a lot of change occurs, a lot of time has passed. At least that is the experience. And that is what rephotography of the environment is capable of doing best, offering a significant passage of time, often compressed, into a form that is to be experienced and then digested. No longer does time conform to a measured standard. No longer is change tied to the inevitable march of the calendar. The pictures challenge the viewer to experience both time and change, and raise questions linking cause and effect. This transformation of fact to awareness is in itself a revolution in offering reasons to care about what is happening to the Earth.

Returning again to Hansen's question regarding how to portray the horror of climate change, environmental photography and the techniques of rephotography offer a merger of facts, poetics and experience. They serve as visual resistance to the debilitating effects of business as usual, ambivalence and helplessness in the face of powerful forces. The photographs tell the stories of different places, and communicate the detailed effects of human activity on the planet. As witnesses to many separate years they also tell us what we stand to lose as a result of change – and most importantly how such losses will feel. The practice calls upon us to understand what once was, to appreciate and hold on to what is left, and to care for the future of wildness.

Notes

1 See the HfE website http://hfe-observatories.org/common-threads/. All photographs, shown in black and white in this chapter, can be seen in full color and at full scale on the HfE website: http://hfe-observatories.org/projects/environmental-rephotography/.
2 For information about the photographic methods used to make rephotographs see Klett 'Rephotography as a Research Method' and 'Three Methods of Presenting'.
3 For more information about the methods behind this project and the many features of the project's DVD see Klett 'Rephotography as a Research Method' and 'Three Methods of Presenting'.
4 In 2011, photographers Mark Klett and Byron Wolfe in collaboration with writer Rebecca Solnit began a new project, currently unpublished and tentatively called 'The Place Everyone Knows'.
5 For Holmund's work on glaciers see: http://bolin.su.se/data/svenskaglaciarer.
6 For more on this project see Martinsson, *Arctic Views Passages in Time*.
7 The reproduction of Axel Goës photographs were published in 1865 as a gatefold in Chydenius, *Svenska Expeditionen till Spetsbergen år 1861*.
8 Charles C. Palmer's watercolour is held at the National Archives in England as item 2, 'Appearance of Land round Magdalena Bay, Spitzbergen, taken from Bird Island, 4 June 1818 by C Palmer' under reference number ADM 344/1963 (not online).
9 Salomon August Andrée led the infamous balloon expedition that left Spitsbergen in 1897 only to vanish into the Artcic. They were found 33 years late in 1930 on an island, Kvitøya, northeast of Svalbard. In the fidnings were notes, exposed film

containing 93 photographs from the expedition, equipment and the bodies of the three men. For more about the expedition visit: http://www.grennamuseum.se

Works cited

Banerjee, Subhankar. *Arctic National Wildlife Refuge, Seasons of Life and Land*. Seattle: The Mountaineer Books, 2003. Print.

Beechey, William. A *Voyage Of Discovery Towards The North Pole*. London: Richard Bently, 1843. Web.

Cardenas, Edgar. *Art-Science for Sustainability*. Unpublished Doctoral Dissertation. Tempe: Arizona State University, 2015. Thesis.

Carson, Rachel. *Silent Spring*. Cambridge MA: Riverside Press, 1962. Print.

Chydenius, Karl. *Svenska Expeditionen till Spetsbergen år 1861 utförd under ledning av Otto Torell*. Stockholm: Norstedts, 1865. Print.

Cronon, William. 'The Trouble with Wilderness; or, Getting Back to the Wrong Nature', *Uncommon Ground: Rethinking the Human Place in Nature*. Ed. William Cronon. New York: W. W. Norton & Co., 1995. 69–90. Print.

Hansen, James. *Storms of My Grandchildren: The Truth about the Coming Climate Catastrophe and Our Last Chance to Save Humanity*. London: Bloomsbury Publishing, 2009. Print.

Holmlund Per. *Glaciärer: gnistrande smycken som ännu pryder våra svenska fjäll*. Karlstad: Votum Gullers, 2012. Print.

Klett, Mark. 'Rephotography as a Research Method', *Longitudinal Visual Research*. Eds Eric Margolis and Luc Pauwels. London: Sage Press, 2008. 114–131. Print.

——'Three Methods of Presenting Repeat Photographs: Exploring Digital Media and the Context for the View, Connecting Two or More Photographs in a Continuous Space, and Extending the Information Contained in the Original View', *Repeat Photography: Methods and Applications in the Natural Sciences*. Ed. Robert Webb. Washington: Island Press, 2011. 32–45. Print.

Klett, Mark, Ellen Manchester and JoAnn Verburg. *Second View: The Rephotographic Survey Project*. Albuquerque: The University of New Mexico Press, 1984. Print.

Klett, Mark, Kyle Bajakian, William L. Fox, Michael Marshall, Toshi Ueshina and Byron Wolfe. *Third Views Second Sights: A Rephotographic Survey of the American West*. Santa Fe: Museum of New Mexico Press, 2004. Print.

Lopez, Barry. Endorsement note for Subhankar Banerjee, *Arctic National Wildlife Refuge, Seasons of Life and Land*. Seattle: The Mountaineer Books, 2003. Print.

Malde, H. E. 'Geologic Bench Marks by Terrestrial Photography', *Journal of Research US Geological Survey*. 1, 2 (March–April 1973): 193–206. Print.

Martens, Friedrich. *Spitzbergische oder Groenlandische Reise Beschreibung, gethan im Jahr 1671. Aus eigner Erfahrunge beschrieben, die dazu erforderte Figuren nach dem Leben selbst abgerissen, (so hierbey in Kupffer zu sehen) und jetzo durch den Druck mitgetheilet*, Hamburg: Gottfried Schultze, 1675. For an English version see: Friedrich Martens, 'Voyage into Spitzbergen and Greenland', in John Narborough, Jasmen Tasman, John Wood and Frederick Marten (sic) *An Account of several late Voyages & Discoveries to the South and North. Towards The Streights of Magelan, the South Seas, the vast Tracts of Land beyond Hollandia Nova, &c. Also Towards Nova Zembla, Greenland or Spitsberg, Groynland or Engrondland, &c*. London: Royal Society, 1694. Web.

Martinsson, Tyrone. *Arctic Views Passages in Time*. Stockholm: Art and Theory Publishing, 2015. Print.

Muir, John, My *First Sumer in the Sierra*. Originally published by Houghton Miffin Co.: Boston, 1911. Edition used in *John Muir, The Eight Wilderness-Discovery Books*, London: Diadem Books, 2004.

Phipps, Constantine, John. A *Voyage Towards the North Pole, Undertaken by His Majesty's Command 1773*, London: J. Nourse. 1774. Web.

Porter, Eliot. *The Place No One Knew*. San Francisco: Sierra Club, 1963.

'The Wilderness Act', Cong. Public Law 88–577 (16 U.S. C. 1131–1136). 88th Congress, Second Session. The United States Government Printing Office. 3 September 1964. Web.

10 Integral ecology in the Pope's environmental encyclical

Implications for environmental humanities

Michael E. Zimmerman

According to some scientists, humankind's current influence on the biosphere has become so extensive that we have entered a new geological epoch, the Anthropocene. The new term is useful for gaining media attention, but concerns about humanity's influence on the biosphere have been growing for decades. At first, academic research about environmentally damaging human activity was largely confined to certain branches of natural science. In the late 1970s and 1980s, however, a few philosophers began to argue that their discipline could shed light on environmental issues and recommend how to address them. Resistance to the very idea of 'environmental ethics' was widespread among professional philosophers, who regarded as absurd the idea that trees or even biomes could have moral standing. Like other humanists, philosophers still sharply distinguished between mind and body, freedom and necessity, historical *human* processes and merely *natural* events. Although accepting Darwin's hypothesis that humans arose from the same evolutionary processes that shaped the rest of life on Earth, humanists typically had little interest in nature as studied by the natural sciences. The humanities were devoted to studying and assessing *human* products, such as texts and works of art, and human events, including those deemed significant in human history.

Expanding the scope of the humanities

New generations of humanists, however, including those who helped to develop environmental literary studies and criticism in the 1990s, increasingly recognised that the traditional humanity vs. nature dualism was no longer tenable for many reasons, including the prospect that human activity would lead to an environmentally grim future. Humanists of various stripes began asserting that their fields offer important perspectives for making sense of and improving the humanity–nature relationship. Leaving environmental research primarily to the natural and social sciences was no longer viewed as a viable option. Increasingly, scientists, policy makers and humanists alike agreed that environmental issues are enormously complex, exhibiting many different facets that are often visible only from the perspectives at work in various methodologies. Integrative or integral approaches to characterising and solving environmental problems were called for. Multidisciplinary and trans-disciplinary research modalities were taken more seriously than ever before, although such modalities have a long way to go before reaching their potential (Brown et al.).

In the eighteenth century, Alexander Pope famously asserted, 'The proper study of Mankind is Man' (1734). Perhaps so, but today, we would insist on a gender-inclusive term such as 'Humankind'. Moreover, we would want 'Humankind' to include the whole array of humans, including the differently abled ones and those not of European descent. Finally, we would acknowledge that Humankind arises from and is thus dependent on material and organic processes that exist 'outside the text'. The gifts of complex language and opposable thumbs, which do set us apart from other (known) beings, let us disclose the structures of the world in ways not otherwise possible. If the proper study of Humankind is Humankind, however, humanists should understand how natural and social sciences depict Humankind's origins, organic constitution, social organisations and future prospects. Humanists can no longer restrict their research to topics traditionally found in departments such as literature, philosophy, history, religious studies and cultural anthropology. Environmental humanists willing to step outside their academic comfort zones can help overcome the gap between what C.P. Snow called the 'two cultures' of the humanities and the sciences (see Castree; Adamson).

In *Integral Ecology: Uniting Multiple Perspectives on the Natural World* (henceforth IE), Sean Esbjörn-Hargens and I take such a step (Esbjörn-Hargens and Zimmerman). We describe a multi-perspectival, developmental approach to characterising and proposing solutions to environmental issues. According to our particular version of 'integral ecology,' findings from the natural sciences and engineering offer important insight into environmental issues, but relying solely on technical assessments for policy decisions amounts to technocracy, the authoritarian rule of experts. A version of this ideology was widely used during the 1950s and 1960s to legitimate the imposition of freeways that destroyed urban neighbourhoods of minorities or others who had no voice in such decisions.

In opposition to such practices, there arose the idea of stakeholder meetings, in which individuals share their *first-person* concerns about projects or environmental issues that affect them. Likewise, *collective cultural* considerations – religious, philosophical, cultural, communal – came to be seen as increasingly important to groups weighing the advantages and disadvantages of a project, such as a dam that would affect long-standing access to and aesthetic appreciation of a free flowing river. Very important as well are *social and natural science perspectives* that reveal systemic problems and opportunities that would otherwise remain hidden from view.

According to IE, all phenomena have interior as well as exterior aspects, and are also individuals as well as members of larger organisations. For example, a bird has some measure of subjective awareness, is constituted by an organic body, seeks to preserve itself as an organism and is also a member of family and flock. In the human domain, IE uses the term 'culture' to refer to shared values and norms (religious, ethical and aesthetic), while 'society' is used to refer to publicly observable behaviour of human social organisations, which social science can examine without any necessary reference to 'interiority' or shared values.

Several other authors who offer integrative worldviews agree that environmental problems must be understood in terms of first-person, cultural and social/natural science perspectives. For example, Felix Guattari maintains that there are three ecologies, social, mental and environmental, all of which are interrelated in many different ways. (Guattari). Holmes Rolston III (*Environmental Ethics*), Edgar Morin (*On Complexity*), Thomas Berry, Brian Swimme and Mary Evelyn Tucker, among others, assert that integrative, multi-perspectival worldviews are important for situating human history within the more-encompassing history of terrestrial and cosmic evolution.

In addition to describing humankind as a product of terrestrial evolutionary processes, IE argues – more so than most other integrative worldviews – that human history also reveals waves of *cultural evolution* or *cultural development*. Today, the idea of cultural development is often the object of suspicion, given that supposedly more 'developed' Europeans and Americans assumed that they were justified in invading and colonising 'backward' peoples. In the contemporary multicultural, post-colonial context, then, terms like 'development' call to mind exploitative hierarchies that must be resisted. While acknowledging such criticism, I continue to think that developmental models can be useful in making sense of complex cultural arrangement.

For instance, the contemporary American population may be said to exhibit three major developmental 'centres of gravity': traditional, modern and postmodern Green. Members of traditional culture are generally conservative and attend church. Although also modern in some ways, traditionalists resist modern trends that undermine their beliefs, often regarded as stemming from the Bible. Moderns assert that only humans have intrinsic value, and hence that non-human entities have only instrumental value. Insofar as modern science, especially ecology, shows that human practices, including industrial pollution and habitat destruction, can damage human well-being, most moderns are willing (in theory) to curb practices that soil the human nest, as it were. Postmodern Greens typically condemn modernity for its arrogant anthropocentrism. (In what follows, I use 'Green' to refer not to a particular political party, but instead to environmentalisms that challenge modernity's anthropocentrism as well as its grand developmental narratives, often used to justify techno-industrial exploitation of nature by capitalist and communist countries.) Not all environmentalists are postmodern Greens, because there are also traditional as well as modern forms of environmentalism. Moreover, not all postmodernists are environmentalists.

Increasing wealth and productivity associated with economic globalisation show up to moderns as improving the human estate, but to postmodern Greens as harming the biosphere or *Gaia*. Many traditionalists side with moderns about economic development, but some evangelical Christians have begun to criticise people who are poor stewards of Creation. The current debate about anthropogenic climate change (ACC) amounts to a culture war between moderns and Green postmoderns (Zimmerman 'Including and Differentiating'). Many modernists claim that climate science has been corrupted by Greens, who have an environmental agenda to install an UN-sponsored authoritarianism. From this point of view, Greens are like watermelons: Green on the outside, Red on the inside. Modernists argue that whatever happens to the climate, poor countries will cope better if they become wealthier, and for the time being fossil fuels are needed for production of wealth. In contrast, Greens depict climate change skeptics as 'denialists,' committed to profit at any cost and willing to risk potentially widespread environmental damage by promoting unrestrained use of climate-changing fossil fuels.

The recently published papal encyclical, *On Care for Our Common Home* (Pope Francis), devotes an entire chapter to what the Pope calls integral ecology. His version of integral ecology shares the multi-perspectivalism adopted by many integrative ecologies, but differs from IE in downplaying the idea of cultural development. The Pope's intervention is important because it shows that religious concerns can shed light on environmental issues, both with respect to their origin and to their possible resolution. Religion, much to the surprise of social scientists who once predicted that it would be relegated to the dustbin of history, retains an important place in global affairs. Indeed, the great majority of the world's population may be characterised as religiously fundamentalist or traditionalist, a long way from modernity, much less from postmodern Green. Despite heading up a traditional religious institution, the Pope's encyclical is primarily a Green document, although not a wholly postmodern one. Rightly concerned about the plight of the world's poor, the Pope emphasises the link between poverty and ill-advised, exploitative treatment of the natural world.

The press anticipated the encyclical as the Pope's statement on climate change. In fact, however, that issue takes up only a small portion of the lengthy document, which argues that ACC is an important symptom of attitudes and practices of a morally unconstrained global capitalism. The encyclical, which shows the influence of work Brazilian liberation theologian Leonardo Boff (*Ecology and Liberation*; *Cry of the Earth*), can be summarised as follows. Armed with the ideologies of private property rights and anthropocentrism, along with the innovative power of modern techno-science, a relatively small class of individuals has created enormous wealth for some at the expense of the many, including not only humans and but also other members of biotic Creation. The Pope's proposed alternative to planet-destroying capitalism is a Christian, Left–Green polity, which emphasises local, Indigenous values, community-oriented work and products, appropriate care for humans and non-humans, and obedience to God's loving plan. Sobriety, humility, kindness, inclusion and financial generosity – these are the kinds of virtues called for by the eco-theology of Pope Francis, who frequently mentions the Creation-affirming, live-simply attitude of his beloved namesake, St Francis of Assisi. In criticising 'consumerism' as a major cultural impetus for exploiting nature, the Pope adopts a standpoint similar to Buddhists, according to whom craving and ignorance lead people to seek happiness in all the wrong places, mainly in purchasing material goods that go far beyond reasonable needs (Loy).

Although he wrote his encyclical primarily as a pastoral document for Catholics, the Pope knew that his work would be read by people with three different and conflicting centres of gravity: pre-modern/traditional, modern and Green postmodern. Hence, he engages in a delicate balancing act, attempting to reconcile Green values with traditional Christianity, while looking to one version of modernity (socialism) to replace another version (capitalism).

Just as traditional culture remains suspicious of modern views and practices, modernity is suspicious of Green postmodernity. To escape from the orbit of pre-modern traditionalism, moderns usually depicted traditional religious beliefs and norms as superstitious nonsense. Instead of integrating the valid aspects of traditionalism, then, moderns *dissociated* themselves from it. Likewise, Greens usually depict modernity's progressive aspirations in the least charitable fashion, thus tending to dissociate themselves from modernity. For Green consciousness, 'nature' means the inherently valuable source of all phenomena, including humankind. Contrary to theocentric biblical traditions and to anthropocentric modernity, so argue the Greens, there is no cosmic hierarchy atop which humankind stands. Human arrogance must be replaced by deep and abiding appreciation and respect for nature.

The importance of evolutionary cosmology

As a Catholic environmentalist, I find much to affirm in the papal document. That we have a Green Pope is remarkable, even though his encyclical offers few if any novel proposals within Christian eco-theology, which has been a going concern for four decades (Northcutt). In line with his preference for consultation, Pope Francis has made liberal use of Church documents and position papers, many of them prepared by the Council of Bishops in one country or another. As a result, the encyclical sometimes reads as if composed by committee. In contrast, the brilliant Catholic theologian Rosemary Radford Ruether's book, *Gaia and God: Ecofeminist Theology of Earth Healing*, covers much of the same ground as the encyclical, but speaks in a single voice and offers a more coherent historical narrative (Ruether). Moreover, Ruether covers important topics overlooked by the encyclical, including the connection between exploiting nature and oppressing women. This omission is not surprising in view of the Church's patriarchal attitude, one of the most visible indications of the Church's resistance to modern norms of gender equity.

Another topic that Ruether explores, but that is barely touched upon by the encyclical, concerns evolutionary theory as it applies to cosmic development. The Church long opposed the idea – proposed in the mid-nineteenth century by Charles Darwin (and Alfred Russel Wallace) – that all life has evolved, even though this remarkable idea has become one of the cornerstones of many branches of natural science. Only in 1950 did Pope Pius XII affirm that there is no 'friction' between evolutionary theory and Church doctrine. Today, Pope Francis emphasises more than any predecessor the extent to which humankind must be understood as a part of nature, not as something alien to or outside of it. By asserting that humankind is a creature dependent on the other creatures constituting the biosphere, the encyclical renounces Neo-Platonic otherworldliness.

The cosmological implications of evolutionary theory were first explored by Georges Lemaître, a Catholic priest from Belgium, who in the late 1920s began publishing papers outlining what would later be called the Big Bang theory. According to this theory, the universe had a beginning and has evolved over billions of years, rather than being eternal and static, as many scientists and philosophers had once supposed. In the late twentieth century, a growing number of cosmologists began devising new universe narratives. Philosopher Steve McIntosh argues that the universe is a generative matrix that – over the course of 13.8 billion years since the Big Bang – has brought forth sequentially matter, life, sentience, and self-consciousness. (McIntosh). As an alternative to twentieth century existentialism, which despaired over a meaningless and thus absurd cosmos, new cosmic narratives ask about the possibly deep significance of the fact that the universe gives rise to beings like us, capable of asking about 'meaning' in the first place. (See Boff *Cry of the Earth*; Chaisson; Davies; Delio; Swimme *The Universe Story*; Swimme and Tucker; Wilber, *Sex, Ecology, Spirituality*; Berry and Tucker.) According to process philosophers and theologians, such as Alfred North Whitehead and Charles Hartshorne, cosmic history is the process by which God arrives at full self-understanding and self-realisation.

Holmes Rolston III offers compelling accounts of cosmic evolution in its material, biological, sentient and self-conscious dimensions (*Environmental Ethics; Three Big Bangs*) Deploying an evolutionary scheme with a hierarchical component, Rolston argues that everything (including the material constituents of the cosmos) has a basic level of inherent value, but that value grows more significant as phenomena evolve, as in the case of terrestrial life, sentience and consciousness. Humans have greater inherent worth than do organisms and species, although humankind is dependent on the non-human world in all its complex levels. Although ascribing great significance to humankind, Rolston makes clear that humankind has arisen from and thus is indebted to not only to other terrestrial life forms and planet Earth itself, but also to the cosmic creativity that made it all possible (Zimmerman 'Integral Ecology's Debt').

A close reading of Rolston's text reveals developmental theistic views, aspects of which are also to be found in integral thinkers such as G.W.F. Hegel, Whitehead, Hartshorne, Teilhard de Chardin and integral theorist Ken Wilber, among others, who have extended the idea of evolutionary development to culture. More so than does Rolston, IE emphasises *human cultural evolution*. The Pope does not much explore cultural evolution, but instead focuses on the need to include multiple perspectives – scientific, political, economic, cultural and individual – in depicting and finding resolutions to complex environmental problems. The Pope writes:

> Recognizing the reasons why a given area is polluted requires a study of the workings of society, its economy, its behavior patterns, and the ways it grasps

> reality. Given the scale of change, it is no longer possible to find a specific, discrete answer for each part of the problem. It is essential to seek comprehensive solutions that consider the interactions within natural systems themselves and with social systems.
>
> (Pope Francis 139)

Papal and postmodern Green resistance to the idea of cultural evolution

One reason the encyclical handles cultural evolution issue gingerly is that, according to Christian theology, the pivotal event in human history was the incarnation of Jesus Christ 2000 years ago. Even if development in consciousness and culture has occurred since then, the Church tends to regard human nature as mostly unchanged, plagued by a proclivity to sin. Another reason for lack of cultural development discourse is that postmodern Greens regard it as an instance of hierarchical thinking, held responsible for so much human oppression and despoliation of nature. IE acknowledges that during the past 250 years efforts to put into practice modernity's noble world-centric ideals, including universal human rights, have often been marred by residual racism, ethnocentrism and imperial power motives, all of which have been rightly criticised by post-colonial theorists and activists. Failure to live up to its own ideals, indeed using those ideals as an excuse to colonise 'un-civilised' lands, reveals one of modernity's dark sides. Another limitation of modernity is its treatment of nature solely as raw material suitable for enhancing human power, and all too often only the power of privileged human classes.

Despite flaws in its conception of 'universal' human traits (all too often defined in Euro-centric terms) and in its application of ideals such as political self-determination, individual liberty, economic well-being, freedom of religion, of the press and so on, modernity has long been a beacon for people who resent the constraints imposed by pre-modern or traditional ways of life. The democratic impulse at work in the recent Arab Spring revolutions was modernist in flavour, even though those promoting the revolutions were aware of modernity's complicity in nineteenth and twentieth century European colonial interventions in North African affairs. The revolutions faltered in part because of resistance from those committed to traditional, pre-modern religious and socio-cultural arrangements. Of course, to defend and adequately to re-conceptualise the ideal of cultural development would require a far longer discourse than is possible within the limits of this chapter.

This encyclical's tendency to minimise cultural evolution is regrettable, but not surprising. For centuries, the Roman Catholic Church stood as a bulwark against many of modernity's new values, including freedom of inquiry, freedom of individual conscience, freedom to elect government authority, and so on. These values arose in part from theological and political discussions that eventually led to the Protestant Reformation. According to Martin Luther, for example, priests are not necessary as intermediaries with God. Instead, every person has a direct relationship to God, who is no respecter of a person's status. The increased standing of individual persons eventually influenced the rise of seventeenth and eighteenth century political reform and revolution. Although Catholics made some important contributions to the rise of democratic principles, Church prelates more often supported aristocracies that were eventually overthrown by popular revolution. Through the nineteenth and much of the twentieth century, the Church regarded modernity (including natural science) with grave suspicion, as an illegitimate rebellion against divine (and papal) authority. Only after World War II did the Church, in Vatican II, find a way to reconcile its teachings with important modern ideals, including individual rights and freedoms. Modern Western culture had evolved beyond the Church's pre-modern attitudes, which then had to play catch-up in order to retain some credibility in the eyes of moderns (and postmoderns) who usually regard religion as a collection of superstitions that support authoritarian regimes.

In the 1960s, just as the Church was trying to come to terms with important modern values, Green postmodernism began to introduce new values of its own. As in the case of modernity, the Church's interest in Green postmodern values was prompted not by its own theological reflections, but rather by external cultural, social, and political events, some of which have provoked at least as much ecclesiastical reaction as did the advent of modern values. The Green movement posited a whole new domain of value – that of the non-human world – that had largely been ignored by moderns as well as by Church members.

Throughout the nineteenth century, a growing number of Western people regarded 'progress' in favourable terms, and tended to discount its destructive consequences as inevitable growth pains. During the twentieth century, however, the dark side of modernity became increasingly evident not only to Westerners, but also to people whose lands had been colonised. World wars conducted using the latest high-tech armaments, grand narratives justifying whatever means were necessary (including the Soviet *gulag*) to carry out ideological aims, environmental pollution, habitat destruction, annihilation of traditional beliefs and practices, and modernity's on-going failure to live up to its own ideals – all these provoked criticism and a search for alternatives.

Green postmoderns concluded that anthropocentric modernity, abetted by natural science and technology and by economies committed to limitless growth, treat nature as nothing more than the raw material needed for enhancing human power and security. Many radical Greens insist that rather than adhering to anthropocentric modernity, we should adopt a dramatic alternative, *biocentrism*, according to which humans are just one more branch on the great bush of life, rather than being an exceptional species at the tip of the evolutionary tree. (Devall and Sessions; Manes). Wisely, the Pope acknowledges that neither an arrogant anthropocentrism nor an anti-hierarchical biocentrism is adequate: there *is* something special about humans, *and* other creatures have goods of their own that call for respect and admiration on the part of humankind, which is both related to and dependent on those creatures.

The human roots of the ecological crisis

In chapter four, 'The Human Roots of the Ecological Crisis,' the Pope's encyclical intends to refute historian Lynn White, Jr.'s influential (but unmentioned) essay, 'The Religious Roots of the Ecologic Crisis' (White), according to which Christianity is the world's most anthropocentric religion. The Pope does his best to meet the challenge posed by White and others, namely, that modern anthropocentrism is in effect a secularised version of themes drawn from Christianity and thus to some extent from Judaism. In *Genesis* we read that humankind is given 'dominion' (a Hebrew term that can mean domination) over Creation. The Pope himself admits that Judaism and Christianity 'disenchanted' Creation by depriving it of divine status: Nature is not God. The Protestant Reformation in particular sought to rid the world of all remnants of 'magical' or 'animistic' thinking, including the Catholic claim that consecrated bread and wine contain the 'real' body and blood of Jesus Christ. Reformation England played a key role in the scientific depiction of nature as mere matter-in-motion. Perhaps it is no accident that England also played a central role in the rise of modern capitalism.

Granted that the Church cannot countenance instances of pantheism, such as *Gaia*-worship, how to represent the relation of Creator to Creation? This question has long vexed monotheistic theologians. Eastern Orthodox Christianity, influenced by its emphasis on the human face of Jesus Christ, rather than restricting itself solely to Scripture as in the Christian West, has explored the Creator–Creation relationship in a way that deserves study by all concerned about this matter (Bartholomew; Foltz). Contemporary theologians often use the concept of *panentheism*, in order to explain that God is present in the world, but also transcends it. God, in other words, cannot be equated either with Creation or with nature. In recent decades, academics from university religious studies departments have persuasively argued that monotheism (Jewish, Christian and Muslim) valourises Creation, thereby imposing religious and moral constraints on how humans may treat it.

Understandably, the Pope hesitates to depict modernity and postmodernity as evolutionary advances *beyond* Christianity, which for him must be the ultimate basis of value and human purpose. Hence, he tries to show that that seeds of modernist values (such as 'authentic human development') and Green postmodern values are prefigured in Holy Scripture. In fact, however, only such concepts are largely missing in the New Testament. Alluding to St Francis, the Pope remarks that today's daunting social and environmental problems:

> have caused sister earth, along with all the abandoned of our world, to cry out, pleading that we take another course. Never have we so hurt and mistreated our common home as we have in the last two hundred years. Yet we are called to be instruments of God our Father, so that our planet might be what he desired when he created it and correspond with his plan for peace, beauty and fullness. The problem is that we still lack the culture needed to confront this crisis
>
> (Pope Francis 53)

To talk about contemporary environmental and social problems, the Pope must move beyond Scripture to modern and postmodern discourses. The following passage from the encyclical draws heavily on German philosopher Martin Heidegger and his student Herbert Marcuse, whose book *One-Dimensional Man* (Marcuse) played a significant role in the rise of Green postmodernism:

> The basic problem goes even deeper: it is the way that humanity has taken up technology and its development according to an undifferentiated and *one-dimensional* paradigm [my emphasis]. This paradigm exalts the concept of a subject who, using logical and rational procedures, progressively approaches and gains control over an external object. This subject makes every effort to establish the scientific and experimental method, which in itself is already a technique of possession, mastery and transformation. It is as if the subject were to find itself in the presence of something formless, completely open to manipulation. Men and women have constantly intervened in nature, but for a long time this meant being in tune with and respecting the possibilities offered by the things themselves. It was a matter of receiving what nature itself allowed, as if from its own hand. Now, by contrast, we are the ones to lay our hands on things, attempting to extract everything possible from them while frequently ignoring or forgetting the reality in front of us
>
> (Pope Francis 106)

Heidegger's anti-modernism had a major influence on French poststructuralists such as Jacques Derrida and Michel Foucault, whose work shaped literary theory, post-colonial discourse, multiculturalism and Green postmodernism. Heidegger claimed that modernity – capitalist and communist alike – is gripped by an insatiable desire for ever greater power over nature, including power over humankind, now regarded as the most important raw material (Heidegger). Starting in the 1970s I began depicting Heidegger's thinking as a foundation for radical Green movements such as Deep Ecology (Zimmerman 'Heidegger on Nihilism'; 'Heidegger and Marcuse'; 'Marx and Heidegger'; 'Toward a Heideggerian *Ethos*'). In the late 1980s, however, I discovered that Heidegger's anti-modernism had much in common with the anti-modernism of National Socialism, to which he lent his public support (Zimmerman *Heidegger's Confrontation*; 'Rethinking the Heidegger'; *Contesting Earth's Future*). One reason Heidegger proved to be such a good fit with Deep Ecology was that they shared attitudes that were not merely postmodern, but at times *anti*-modern. Some Deep Ecology discourse, so I concluded, was perhaps unwittingly reminiscent of eco-fascism (Zimmerman 'The Threat of Ecofascism'). Ken Wilber's works, however, offer a way to *differentiate* between modernity's noble contributions and its limitations, without adopting a totalising anti-modernistic stance (*Up from Eden*; *Sex, Ecology, Spirituality*). Instead of dissociating itself from modernity, a healthy Green postmodernism should explicitly acknowledge and integrate what is worthy about modernity, while criticising and transcending what is problematic about it. Such a move could in turn allow moderns to take Green values more seriously.

What is the best way forward: capitalism or Green socialism?

In his encyclical the Pope emphasises the Judeo-Christian imperative to care for and to protect the poor, oppressed, and dispossessed, especially those in developing countries who are put at even greater risk by predatory global market forces. He calls for a redistribution of technology and wealth as a way to address practices that have exploited the weak in the 'winner take all' mentality of the market system. Although denying that he is calling for a 'return to the Stone Age,' the Pope insists that there is nothing 'magical' about the market system that enables it to solve all environmental and social problems, if governments and other social institutions simply get out of the way. The Pope recommends what amounts to a socialist alternative to capitalism.

IE maintains that the *values* at work in capitalism and communism shape their *economic* principles. Liberal capitalism emphasises the link between individual liberty and free markets, whereas communism insists that the value embodied in universal material well-being is a necessary condition for healthy individuation. Capitalism and communism, however, accord only *instrumental* value to nature. Whereas capitalism calls for exploiting nature to serve *individual* human ends, socialism exploited nature for *collective* human ends. Marx praised profit-motivated capitalism for making possible the machine technology needed to end material scarcity, thus eliminating alienated labour and allowing for full human development. For Marx, the next phase in social evolution is collective ownership of the means of production, without which individualist ideology would continue to justify exploitation of labour, along with vast disparities in wealth, opportunity, and social status.

Although both systems have been responsible for great environmental harm, Soviet and Communist Marxism may have done even worse damage, in part because authoritarian communism limited press freedom, NGOs, and political debate far more than does liberal capitalism (see Feshbach; Shapiro).

Individualism and communalism are enduring value polarities that ideally can generate a healthy, creative tension. The prospect of total victory by either polarity should give us pause. Neither social atomism, in which community vanishes, nor all-embracing collectivism, in which there is no room for individuality, is a desirable option. Capitalism and communism are *ideologies*, political–philosophical viewpoints that reveal some important truths, while simultaneously concealing other truths, especially those that prove inconvenient to or contradict the ideology in question. Capitalist criticisms of communism, and communist criticisms of capitalism, are happy to show the shortcomings of each other, but rarely engage in genuine self-criticism. Such criticism would reveal that that the opposing ideology contains legitimate values that are underrepresented in one's own ideology. An academic humanist interested in discovering the values embodied in modern and postmodern ideologies ought not to adhere rigidly to one particular ideology; otherwise, critique of opposing ideologies will amount to knocking over a straw man. A humanist should (arguably) engage in Socratic self-examination, even though doing so may prove uncomfortable.

Capitalism generates unparalleled prosperity, but has many problems as well, including boom and bust cycles, great differences in wealth and status, exploitation of nature as well as of human beings, and so on. Capitalism's defenders are right in saying that such unfortunate distortions often result from government interference in the workings of markets, but critics point out that market economists are clueless about how to decrease the current growth in income disparity. Can the challenges of widespread poverty and hunger in the USA, as well as a shrinking middle class, really be resolved solely by eliminating government restrictions on market practices? There are good reasons to answer to offer a negative answer. In fact, markets have always been limited by cultural, moral, political and legal provisions. The papal encyclical reminds us that despite the enormous material wealth generated by capitalism, millions of people languish in poverty even in capitalist countries.

Twenty-first century social and environmental problems, however, are best described as 'wicked,' not only because of their great complexity, but also because efforts to solve one or another such problem often generate unanticipated negative consequences (see Hulme; Brown et al.). Government restrictions aimed at a perceived environmental problem sometimes made matters worse. In recent decades, environmentalists have sometimes found ways to enlist market forces to achieve environmentally laudable aims (Anderson and Leal). Although to some such free market environmentalism may seem like an oxymoron, in fact there are market entrepreneurs who want to achieve both human well-being and nature protection by using market mechanisms. Free market environmentalists are not alone in saying that significant concern about the natural environment, including industrial pollution and habitat destruction, grows only *after* people have achieved a certain level of wealth. In its headlong rush to catch up with modern economic development, for example, China grew rich at the expense of its natural environment. Today, however, the newly minted middle class increasingly demands alternatives to water and air pollution that threaten millions with early death.

Markets can help people escape from poverty, but success in such matters is not easy. For one thing, markets work best when a society operates at something approaching the modern centre of gravity, which *ideally* demands transparency in public economic transactions. How to encourage modern economic/political conditions (including transparency, the rule of law, constitutional democracy, etc.) necessary for homegrown wealth in countries where traditional values and oligarchy stand in the way? Moving from a traditional or pre-modern to a preferably indigenous modern centre of gravity takes time, effort, struggle, investment, education, and luck. There is no guarantee that it will occur at all, although in the past two centuries the drift in human societies has been from pre-modern to modern, and then to Green postmodern. Culture is a key factor often underappreciated by institutions such as the World Bank in making loans to countries with extensive poverty. A traditional country has its own kind of economy, governed by traditional attitudes, beliefs, and practices that will not vanish simply because a cadre of bureaucrats proposes to conduct business in a very different way. Attempts to resolve complex issues must be carried out with care and with deference to traditional *mores*. Imposing 'modernity' on traditional peoples is one reason that they are so suspicious of it.

The Pope is rightly concerned about the poor, but poverty as a percentage of world population has been decreasing for decades. Hundreds of millions of people have risen into the middle class, a group that the Pope concedes that he overlooked while focusing on the poor (Yardley). With growing wealth comes the opportunity to purchase more things than one really needs. The digitally-transformed culture industry shapes and stokes consumer desire, in part by gleaning personal information from vast digital networks. Exhortation alone, however, is unlikely to change the human appetite for nicer clothes, cars, houses, food, as well as for more frequent travel. The risk of consumerism, however, is no argument against creating conditions that allow people to climb out of poverty.

Few contemporaries would argue that human flourishing is possible without access to electricity. Today, however, 1.5 billion people have no electricity at all. How to provide them with it without exacerbating ACC? Here the clash between moderns committed to human development, on the one hand, and Greens concerned about ACC, on the other, becomes most pointed. The Pope's encyclical observes that climate change – whatever its causes – would affect poor countries the most.

This observation must be balanced, however, by one arising from a different perspective. Whatever the consequences of climate change might be, countries with greater wealth can prepare for it far better than poor countries. *An Ecomodernist Manifesto* (Asafu-Adjaye et al.) argues that industrial modernity, despite the environmental problems it has caused, can deliver human flourishing while simultaneously preserving habitat for the rest of life on Earth. Rolling back modernity is not an option. Progressive environmentalists suggest that increasing material well-being for humankind both can and ought to be compatible with protecting nature (Zimmerman 'On Reconciling Progressivism'; Nordhaus et al. n. pag.). If Greens are serious about slowing climate change, so argues the *Manifesto*, they must allow for new nuclear power plants, which do not use carbon fuels; likewise, if Greens are serious about protecting habitat, they must stop opposing genetically modified organisms (GMOs), which one day may return farmland to habitats, even while feeding additional millions of people.

Critics of eco-modernisation reply that its proposed remedy is in fact a reprise of capitalist modernity's overconfidence in markets and techno-science to solve all problems. For such critics, only reorganisation of economic and society according to a democratic socialist model provides offers a plausible way forward in the twenty-first century. Some modern and postmodern Greens recommend economic de-growth to curb the consumerism held responsible for environmental damage (Caradonna, Jeremy, et al.). The most effective way to slow the global economic engine would be to dramatically decrease fossil fuel use. Many moderns insist, however, that abandoning fossil fuels without alternative energy supplies in place would cause a global economic contraction that would lead countless millions to fall back into poverty.

Moderns often read the anti-market, anti-economic development positions promoted by Greens as a version of anti-humanism. To encourage more moderns to embrace Green concern for the well-being of the natural environment, Greens – as the leading edge of cultural evolution – must demonstrate their commitment to human advancement. That commitment is present in the encyclical, but the Pope's understanding of how to define and achieve such advancement remains limited. Were the encyclical to distinguish more carefully between differing levels of cultural and psychological development, the document may have been able to draw more acute distinctions and to offer more effective recommendations.

Concluding remarks

In this chapter, I have shown that the recent papal encyclical shares IE's multi-perspectival, inclusionary approach to defining and resolving environmental problems in a way that is consistent with modernity's goal of enhancing human well-being. Those working in the environmental humanities should take special note of the Pope's emphasis on the role played by *culture* in social, economic, political, and environmental problems and opportunities. Although economic calculations aimed at enhancing individual welfare are important, they cannot represent the full scope of environmental values (Sagoff). Just as humanists must become more deeply acquainted with findings from the natural and social sciences, representatives of the latter fields must also recognise that understanding human aspirations and Green values is crucial to creating a world that works for humanity and nature alike.

The Pope maintains that a Green sensibility is necessary for achieving social justice. Although other popes have spoken in favour of environmental aims, Pope Francis is the first to foreground humanity's interdependence with the rest of terrestrial nature. Given that it was only during Vatican II (1962–1965) that the Church sought to accommodate itself to many important goals of modernity, an embrace of Green values only fifty years later is something to applaud. The nobility of Green lies in expanding the domain of moral standing beyond the human, so as to include all life and the biosphere itself. The challenge for Greens, including (apparently) the Pope is how to integrate the worthy values of modernity even while continuing to criticise its dark side, including its lack of respect for the natural world. As one transcends a given phase of cultural development, there is a tendency to dissociate oneself from the prior stage. Had the Pope included more discussion of cultural development, and the dangers of dissociation from prior phases of cultural development, he would have gone even further in promoting inclusive dialogues needed to address challenges posed by global poverty and environmentally unsustainable practices.

Works cited

Adamson, Joni. 'Humanities', *Keywords for Environmental Studies*. Eds. Joni Adamson, William Gleason and David N. Pellow. New York: New York University Press, 2015. 135–139. Print.

Anderson, Terry L. and Donald R. Leal. *Free Market Environmentalism for the Next Generation*. New York: Palgrave MacMillan, 2015. Print.

Asafu-Adjaye, John, Linus Blomqvist, Stewart Brand, Barry Brook, Ruth Defries, Erle Ellis, Christopher Foreman, David Keith, Martin Lewis, Mark Lynas, Ted Nordhaus, Roger Pielke, Jr., Rachel Pritzker, Pamela Ronald, Joyashree Roy, Mark Sagoff, Michael Shellenberger, Robert Stone, Peter Teague. *An Ecomodernist Manifesto*, 2015. Web.

Bartholomew, Ecumenical Patriarch. *Toward an Ecology of Transfiguration: Orthodox Christian Perspectives on Environment, Nature, and Creation*. Ed. John Chryssavgis. New York: Fordham University Press, 2013. Print.

Berry, Thomas. *The Dream of the Earth*. San Francisco: Sierra Club Books, 1998. Print.

Berry, Thomas and Mary Evelyn Tucker. *The Sacred Universe: Earth, Spirituality, and Religion in the Twenty-first Century*. New York: Columbia University Press, 2009. Print.

Boff, Leonardo. *Cry of the Earth, Cry of the Poor*. Maryknoll, NY: Orbis Books, 1997. Print.

——*Ecology and Liberation: A New Paradigm*. Maryknoll, NY: Orbis Books, 1995. Print.

Brown, Valerie A., Joan A. Harris and Jacqueline Russell, *Tackling Wicked Problems: Through the Transdisciplinary Imagination*. New York: Routledge, 2010. Print.

Caradonna, Jeremy, Iris Borowy, Tom Green, Peter A. Victor, Maurie Cohen, Andrew Gow, Anna Ignatyeva, Matthias Schmelzer, Philip Vergragt, Josefin Wangel, Jessica Dempsey, Robert Orzanna, Sylvia Lorek, Julian Axmann, Rob Duncan, Richard B. Norgaard, Halina S. Brown and Richard Heinberg. *A Degrowth Response to an Ecomodernist Manifesto. Resilience*, 2015. Web.

Castree, Noel. 'The Anthropocene and the Environmental Humanities: Extending the Conversation', *Environmental Humanities* 5 (2014): 233–260. Print.

Chaisson, Eric. *Cosmic Evolution: The Rise of Complexity in Nature*. Cambridge: Harvard University Press, 2002. Print.

Davies, Paul. *Cosmic Jackpot: Why Our Universe is Just Right for Life*. New York: Houghton Mifflin Company, 2007. Print.

De Chardin, Teilhard. *The Phenomenon of Man*. New York: Harper Perennial Modern Classics, 2008 [1955]. Print.

Delio, Ilia. *Making All Things New: Catholicity, Cosmology, Consciousness (Catholicity in an Evolving Universe)*. New York: Orbis Books, 2015. Print.

Devall, Bill and George Sessions. *Deep Ecology: Living as if Nature Mattered*. Layton, Utah: Gibbs Smith Publishers, 1991. Print.

Esbjörn-Hargens, Sean and Michael E. Zimmerman. *Integral Ecology: Uniting Multiple Perspectives on the Natural World*. Boston: Shambhala, 2009. Print.

Feshbach, Murray. *Ecocide in the USSR: Health and Nature under Siege*. New York: Basic Books, 1993. Print.

Foltz, Bruce V. *The Noetics of Nature: Environmental Philosophy and the Holy Beauty of the Visible*. New York: Fordham University Press, 2013. Print.

Hartshorne, Charles. *Natural Theology for Our Time*. Chicago: Open Court Publishing Company, 1973. Print.

Hegel, G.W.F. *Phenomenology of Spirit*. Trans. A.V. Miller. New York: Oxford University Press, 1976. Print.

Heidegger, Martin. *The Question Concerning Technology and Other Essays*. Trans. William Lovitt. New York: Harper Torchbooks, 1977. Print.

Hulme, Mike. *Why We Disagree about Climate Change: Understanding Controversy, Inaction and Opportunity*. Cambridge and New York: Cambridge University Press, 2009. Print.

Guattari, Felix. *The Three Ecologies*. Trans. Ian Pindar. New York and London: Bloomsbury Academic, 2008. Print.

Loy, David R. *A New Buddhist Path: Enlightenment, Evolution, and Ethics in the Modern World*. Somerville, MA: Wisdom Publications, 2015. Print.

Manes, Christopher. *Green Rage: Radical Environmentalism and the Unmaking of Civilization*. Boston: Back Bay Books, 1991. Print.

Marcuse, Herbert. *One-Dimensional Man: Studies in the Ideology of Advanced Industrial Society*. Boston: Beacon Press, 1964. Print.

McIntosh, Steve. *Evolution's Purpose: An Integral Interpretation of the Scientific Story of Our Origins*. New York: SelectBooks, 2012. Print.

Morin, Edgar. *On Complexity (Advances in Systems Theory, Complexity, and the Human Sciences)*. New York: Hampton Press, Inc, 2008. Print.

Nordhaus, Ted, Michael Shellenberger and Jenna Mukuno. 'Ecomodernism and the Anthropocene: Humanity as a Force for Good', *The Breakthrough Journal* 5 (Summer 2015). Web.

Northcutt, Duncan S. 'Ecology and Christian Ethics', *The Cambridge Companion to Christian Ethics*. Ed. Robin Gill. Cambridge: Cambridge University Press, 2001. 209–227. Print.

Pope, Alexander. 1734. *An Essay on Man*. Stockbridge, Massachusetts: Hard Press Editions, 2006. Print.

Pope Francis. *On Care for Our Common Home: Laudato Si*. Washington, DC: United States Conference of Catholic Bishops, 2015. Print.

Rolston, Holmes III. *Environmental Ethics: Duties to and Values in the Natural World*. Philadelphia: Temple University Press, 1989. Print.

——*Three Big Bangs: Matter-Energy, Life, Mind*. New York: Columbia University Press, 2010. Print.

Ruether, Rosemary Radford. *Gaia and God: Ecofeminist Theology of Earth Healing*. New York: HarperOne, 1994. Print.

Sagoff, Mark. *Price, Principle, and the Environment*. Cambridge University Press, 2004. Print.

Shapiro, Judith. *Mao's War against Nature: Politics and Environment in Contemporary China*. New York: Cambridge University Press, 2001. Print.

Swimme, Brian. *The Universe Story: From the Primordial Flaring Forth to the Ecozoic Era – A Celebration of the Unfolding of the Cosmos*. New York: Harper One, 1994. Print.

Swimme, Brian and Mary Evelyn Tucker. *Journey of the Universe*. New Haven: Yale University Press, 2014. Print.

White, Jr., Lynn. 'The historical roots of our ecological crisis', *Science* 155. 3767 (1967): 1203–1207. Print.

Whitehead, Alfred North. *Process and Reality*. Eds David Ray Griffin and Donald W. Sherburne. New York: Free Press, 1978. Print.

Wilber, Ken. *Sex, Ecology, Spirituality*. Boston: Shambhala, 1995. Print.

——*Up From Eden: A Transpersonal View of Human Evolution*. Wheaton, Illinois: Quest Books, 1981. Print.

Yardley, Jim. 'Pope Says He's Overlooked the World's Middle Class', *New York Times*, 13 July 2015. Web.

Zimmerman, Michael E. *Contesting Earth's Future: Radical Ecology and Postmodernity*. Berkeley and Los Angeles: University of California Press, 1994. Print.

——'Heidegger and Marcuse: Technology as Ideology', *Research in Philosophy and Technology* II (1977): 245–261. Print.

——'Heidegger on Nihilism and Technique', *Man and World* VIII (November 1975): 399–414. Print.

——*Heidegger's Confrontation with Modernity: Technology, Politics, Art*. Bloomington: Indiana University Press, 1990. Print.

——'Including and Differentiating among Perspectives: An Integral Approach to Climate Change', *Journal of Integral Theory and Practice* 4.4 (2010): 1–26. Print.

——'Integral Ecology's Debt to Holmes Rolston III', *Integral Ecologies: Nature, Culture, and Knowledge in the Planetary Era*. Ed. Sam Mickey, Sean Kelly, and Adam Robert. Albany: SUNY Press, forthcoming. Print.

——'Marx and Heidegger on the Technological Domination of Nature', *Philosophy Today* XXIII (Summer 1979): 99–112. Print.

——'On Reconciling Progressivism and Environmentalism', *Explorations in Environmental Political Theory*. Ed Joel J. Kassiola. Armonk, New York: M.E. Sharpe, Inc., 2003. 149–177. Print.

——'Rethinking the Heidegger – Deep Ecology Relationship', *Environmental Ethics* 15.3 (Fall 1993): 195–224. Print.

——'The Threat of Ecofascism', *Social Theory and Practice* 21 (Summer 1995): 207–238. Print.

——'Toward a Heideggerian *Ethos* for Radical Environmentalism', *Environmental Ethics*, V (Summer 1983): 99–131. Print.

Part III
Country

11 Radiation ecologies, resistance and survivance on Pacific Islands

Albert Wendt's *Black Rainbow* and Syaman Rapongan's *Drifting Dreams on the Ocean*

Hsinya Huang and Syaman Rapongan[1]

This chapter examines radiation ecologies and nuclear colonialism in the trans-Pacific, trans-Indigenous context. This work is illustrative of over a decade of collaboration between academics, scientists and Taiwanese Indigenous artists and writers that has contributed to the development of a school of cultural study that links Taiwanese Indigenous studies to Native North American frameworks of native and Aboriginal cultures.[2] Reading Albert Wendt's novel *Black Rainbow* (1992) and Syaman Rapongan's memoir *Drifting Dreams on the Ocean* (2014), we argue that nuclear colonialism creates an ecological debt to Indigenous islanders. Both writers testify to the nuclear Pacific as an ongoing presence across the Pacific region, which remains in the oceanic body and continually increases the risk of ecological degradation. Both writers reflect on the ecological debt–degradation link through the acts of mourning and writing. Albert Wendt was born into a Samoan family and migrated across the Pacific, from Samoa, to Fiji, to Hawai'i and New Zealand. Inspired by the *Black Rainbow* artwork series of Maori artist Ralph Hotere protesting nuclear tests in the Pacific, his novel *Black Rainbow* 'indigenises' English, in order to raise questions about nuclear colonialism in the Pacific. Syaman Rapongan was born and raised in *Pongso no Tau*, the home island of the Tau people, located 40 kilometres southeast of Taiwan, where men and women used to live a life in accordance with their oceanic traditions, which are part of the larger Austronesian culture. He left Orchid Island for education on the main island of Taiwan when he was a teenager, remaining abroad for decades to work in Taipei as an urban Indigene. There, he participated in the Aboriginal demonstrations of the 1980s, of which the most significant was the Tau-led protest against the storage of nuclear waste on *Pongso no Tau*.

Drawing upon Wendt's and Rapongan's literature respectively written in Sinophone script and English, this chapter explores the permeating effect of nuclear toxicity across the Pacific as the Indigenous writers experience it first-hand. Both writers show how ocean ecologies have been severely impacted by radiation. Wendt's and Rapongan's common objective of (w)righting the nuclear

Pacific serves as a medium for the expression of trans-Indigenous solidarity across the Pacific. Wendt's novel specifically links Indigenous responses to nuclearisation and its assault on Indigenous history between New Zealand and French Polynesia; Rapongan's memoir focuses on the history of the struggle between the Tau people and the Taiwan Government, which mirrors in significant ways the ongoing conflict between Pacific islanders and the Western imperial powers. A trans-Indigenous approach to their literature contributes to the ongoing survival of the Pacific islanders under the nuclear devastating effects, attesting to what Gerald Vizenor calls 'narratives of survivance' (*Survivance* 1). In their anti-nuclear narratives, the Pacific writers are beginning to signify a bioregional site of coalitional and transnational promise as the Indigenous people envision new forms of ecological solidarity and inspire an imagination of co-belonging within cultures of collective dissent.

The nuclear Pacific

In her trilogy essays, 'Radiation Ecologies and the Wars of Light' (2009), 'Heliotropes: Solar Ecologies and Pacific Radiations' (2011) and 'The Myth of Isolates: Ecosystem Ecologies in the Nuclear Pacific' (2013), Elizabeth DeLoughrey suggests that the 'global rise of militarized radiation transformed our relationship to light and that the relationship between light and ecology is more than metaphysical. In fact, ecology as a discipline has close ties to the radioactive militarisation of the Pacific' ('Radiation Ecologies' 472). The field of 'radiation ecology' was established in the Pacific after the U.S. set up the Manhattan Project to develop the first nuclear weapons and dropped two bombs in Hiroshima and Nagasaki in 1945. Since that time, the Pacific has become a region 'often deemed peripheral to modernity and yet the site of nearly continuous nuclear weapons testing,' as DeLoughrey puts it brilliantly ('Heliotropes' 239).[3] This is due to the Western imaginaries of islands being remote, isolated, empty, small and insignificant, apt for any scientific experiments including nuclear testing and dumping of toxic waste; islands have been associated with the 'contained space of a laboratory,' an ecosystem of enclosure, i.e. islands as isolates, which would 'create extreme conditions' and bear obsessive inscriptions of modern scientific discoveries ('The Myth' 167–184). DeLoughrey argues that the Age of Ecology and nuclear colonialism coincide in history, 'as a dual threat to the planet in terms of nuclear annihilation (the explosive event) as well as radiative fallout (the aftermath)' ('Heliotropes' 236). The legacy of nuclear militarism and the atomic age not only continues to impact the island space and its Indigenous inhabitants but it also continues to impact humanity and all global ecologies.

In the face of continuing radioactive abuses, such Pacific critics as Epeli Hau'ofa, Anthony van Fossen, David Robie, David Fischer and others, formulate the history and future of the Pacific islands in terms of apocalyptical perceptions of nuclear affected water-/island-scapes.[4] Contained within these perceptions is environmental destruction and ecological devastation, as captured in van Fossen's resonant phrase 'the ecology of doomsday,' which 'refers both to an apocalyptic

Pacific future and to a continuing record of ecological abuses' in the Pacific, attributable to the military-industrial schemes of the imperial West (van Fossen 40, qtd. in Huggan and Tiffin 54). The most atrocious of these abuses are those practised in the name of 'nuclear colonialism' (Robie 146, qtd. in Huggan and Tiffin 54). The systematic exploitation is enormous. As a result, 'the post-Second World War arms race turned large swathes of the Pacific into a military zone and, more specifically, a nuclear arena, with the mantra of "national security"' being 'invoked to justify sustained nuclear bomb and missile testing' in Micronesia by the U.S. and Britain, and in Polynesia by France (Robie; Fischer, qtd. in Huggan and Tiffin 56). The Indigenous groups of the Pacific trespass the boundaries of nation-states and the colonial mapping of the Pacific and their home islands. Following a long tradition of Pacific migration and fluidity, they are continually impacted by nuclear colonialism on the waters and islands of the Pacific, and the devastating effects of radiation ecologies can hardly be overestimated.

And yet, the critique of nuclear colonialism should not merely stress an abusive past and/or an apocalyptic future; it should also highlight an ongoing process in which Indigenous writers, and anti-nuclear critics and activists, are creating opportunities for constructing a world that will be more just. Interlocking Indigeneity and ecologies in the Pacific should be seen as a stepping stone in the process of decolonising island space and freeing island inhabitants from nuclear consequences. We agree with Graham Huggan and Helen Tiffin's conceptualisation of a 'symbiotic relationship ... between decolonisation and the campaign for a nuclear-free Pacific' (54). The Pacific decolonising agenda reframes the continuance and survival of Pacific islanders' culture in a nuclear space, underlining the significance of anti-nuclear narratives and environmental movements of the Indigenous Pacific (see Hau'ofa). Not only does it emphasise the active presence of the Indigenous subject position in these narratives and movements but it also draws on the solidarity of Indigenous peoples in an 'active repudiation of dominance, tragedy, and victimry' in a nuclear space. In his book *Manifest Manners*, the esteemed Native American writer Gerald Vizenor calls for new narratives of survivance as an ethical act of defending our planet: 'The radioactive ruins and chemical wastes of our time are new worries of our time and without the narratives of regeneration. The winter memories of survivance are denied in the ruins of a chemical civilisation' (120). In the face of the contemporary threat of nuclear blight, Vizenor prescribes no easy solution. We cannot count on the old narratives of regeneration, of the Earth as a nurturing mother, or of natural mediation. Vizenor refutes the victimry or mere survival and endurance in the world which has been abused by nuclear testing and its fallout. Nuclear victimry invokes Indigenous solidarity across the Pacific and 'survivance is an active repudiation of dominance, tragedy, and victimry' (*Fugitive Poses* 15). In this, Vizenor claims the active presence of the Indigenous peoples in solidarity and an act of resistance against dominant nuclear discourse and cultural narratives.

Viznor's novel *Hiroshima Bugi: Atomu* 57 is especially relevant in that it pushes to the forefront the issue of survivance in a nuclear era: 'survivance is a creative, concerted consciousness that does not arise from separation, dominance, or concession nightmares' (18). The novel begins in the ruins of the Atomic Bomb Dome, where the protagonist Ronin creates a new calendar that starts with the first use of atomic weapons, Atomu One. He makes explicit the intimate relationship between the Ainu of Indigenous Japan and the Anishinaabe of Native America in the character Ronin as a 'hafu' (half) or mixed blood, whose father, Nightbreaker, was an Anishinaabe from the White Earth reservation of the U.S. and whose mother might have been an Ainu from Hokkaido (Carson 445). Ronin becomes the medium that connects Indigenous Japan with Native America. The uranium bombs, which were created in native lands and dropped in Hiroshima, fatefully connect people in Asia Pacific with Native Americans in common resistance against nuclear imperialism. Likewise, Leslie Marmon Silko depicts in her *Ceremony* that uranium bombs have enclosed human beings in 'a circle of death' as the U.S. government mined the rocks on the reservations and turned them into atomic bombs which killed numerous civilians in Hiroshima and Nagasaki and left even more sickened by atomic fallouts. The protagonist Tayo becomes overwhelmed by the fact how people across the Pacific are connected by the uranium bombs and finally sees the way how 'all the stories fit together – the old stories, the war stories, their stories – to become the story that was still being told' (246). With diverse cultural and linguistic backgrounds, Indigenous writers across the Pacific have been telling a common story of nuclear plight, Indigenous solidarity and survivance which, we hope, will arouse the consciousness of humanity and revive the planet from the nuclear ruins and reconstruct our fragile environment into a more healthy and just home/ community.

Black Rainbow

Albert Wendt addresses nuclear colonialism and Indigenous survivance in *Black Rainbow* from his unique perspectives as a migrant Pacific islander. Inspired by Ralph Hotere's artwork, he expresses his opposition to the nuclear colonialism of the world's superpowers in the waters and islands of the Pacific. In 1986, Hotere produced his Black Rainbow series as an outright and outrageous response to the French nuclear testing in Moruroa Atoll and to the sinking of the Greenpeace protest ship *Rainbow Warrior* by the French secret service in Auckland in the same year. This series inspired Wendt's novel *Black Rainbow*. Traditionally perceived by the Polynesian and Austronesian seafaring peoples as a symbol of ethnic diversities and natural ties with their ancestral spirits, the 'rainbow' in both Hotere's and Wendt's work transcends literal understanding to incorporate profound meanings and updated concerns of ecological devastation by nuclear colonialism.

The rainbow is rendered as black. Multiple meanings can be associated with blackness. First, it symbolises the explosive events of nuclear testing which take place regularly in the Pacific, turning the Pacific waters, which the colonisers named in terms of its peacefulness, into a bleak and dark world of death. Though Wendt makes few overt references to nuclear colonialism in the Pacific, his frequent imaginaries of the blackness denote degeneration and death and evoke 'a fear of annihilation linked to the ever-present nuclear threat' (Huggan and Tiffin 55). Hotere's artwork makes explicit Wendt's protest against nuclear exploitation in the Pacific. The image of the rainbow resembles the billowing mushroom clouds of Hiroshima and Nagasaki, representing 'the existential void to the thread of nuclear catastrophe' (Huggan and Tiffin 55). This association is implicated in the opening conversations between the narrator and his wife:

> Their home is haunted by a lithograph of a black rainbow with a 'countdown' on it [...] She glowed as she looked at the Hotere lithograph on the wall above my head. *Black Rainbow/Moruroa* [...]?
>
> I paused in front of the Black Rainbow. I looked at it closely. Recognised the thick black arch to be the rainbow. But the numbers, 1 to 14, on either side of the upsurging cloud? The countdown to what?
>
> (9–10)

Counting down suggests the nuclear doomsday clock that frames 'the temporality of the text, and, by extension, the Pacific as a region' (DeLoughrey 'Heliotropes' 242). The clock which stopped at 8:15 a.m. when Little Boy (codename of the atomic bomb) destroyed the city of Hiroshima, on 6 August 1945, was now installed at one of the entrances of Hiroshima Peace Museum as a temporal reminder of the tragic history of the city. Likewise, Hotere's 'Black Rainbow' is a lithograph that functions as a truth-finding device, a symbol and an icon of nuclear colonialism that happens as part of the daily experiences of the Pacific islanders since the first bomb was dropped in Hiroshima. It was 'Summer's ending' (9), as Wendt writes, and any sensitive reader can quickly associate the heliographic symbol with the explosive event of 6 August 1945. Furthermore, blackness also suggests the fallout of the Hiroshima A-bomb after the bombing, which appeared in the form of precipitation, as black rain. This motif of blackness was taken up by Masuji Ibuse (井伏鱒二) in his famous novel of the Japanese atomic-bomb literature (原爆文学, Genbaku bungaku) 黒い雨 (*kuroi ame*, 1966) or *Black Rain* in English translation (1969).

Wendt attests to the significance of nuclear criticism in contemporary discourses and narratives of environmental and ecological studies. The consequences of exposures to radiation for the Pacific islanders and biodiverse island ecologies have been devastating. Wendt's writing to 'right' the nuclear Pacific serves as a medium for the expression of the Pacific solidarity for the civilians, who are the custodians of the largest waters on the planet.[5] Though the novel as a futuristic science fiction depicts the totalitarian world of post-nuclear holocaust New Zealand, Wendt's vision is trans-national as well as trans-Indigenous. The colonialism evinced in *Black Rainbow* is not New Zealand or Maori centred (Bush 187), but rather it extends nuclear/colonial criticisms beyond New Zealand, in this case to French nuclear testing in Polynesia, as Wendt draws his title from Ralph Hotere (Alessioa 265). In an interview with Michael Neill, Wendt describes how Hotere adds to his Black Rainbow series every time the French exploded a nuclear device on Mururoa, which inspires his writing: 'Ralph gave me one of the prints in 1986 and that became the inspiration for the short story "Black Rainbow", which has turned into the novel'.[6] Hotere's art represents events that testify enduring legacy of colonial imperialism and decimation of Indigenous islands and cultures of the Pacific (Ellis 110). By employing Hotere's Black Rainbow as the title, Wendt retrieves a subtext of the trans-Pacific protest against nuclear colonialism.

The Hotere lithograph Black Rainbow is the pivotal theme that Wendt's novel revolves around. In addition to the devastating nuclear tragedies that have haunted the daily life of the Pacific islanders, blackness in Polynesian tradition also stands for something that is redemptive, hence building up a hopeful message to counteract the colonial abuses and misuses of nuclear energy. It refers to the fertile womb from which humans come and to which they will return – 'darkness of creation and possibility' (Sharrad 207). The narrator's wife uses it to call to the spirits of the original inhabitants of Maungakiekie, thus immediately evoking the presence of the original inhabitants:

> For me Maungakiekie was one because it was there that my wife, with her courage and sight, had started our rebellion against the Tribunal. She'd summoned the agaga [spirit] of our ancient Dead with the Hotere icon to hold back the doomsday clock; she'd linked us again to the earth and our Dead.
>
> (242)

Maungakiekie is the original Maori name of One Tree Hill (Auckland), the place where the tōtara tree native to New Zealand grew and was used in the making and carving of waka, which is the Maori name for canoes that started to sail across the Pacific Ocean 2000 years ago before Western settlers arrived. The narrator meditated on 'One Tree Hill and how it overshadowed [his] neighborhood; I preferred its original Maori name, Maungakiekie, and its phallic obelisk memorial to the Maori tribe who'd once occupied the hill as a fortified pa' (10). The word 'pa' can refer to any Maori fortified village or defensive settlement. As a verb, 'pa' also means 'to participate, act together, act in concert, join in (an undertaking)' (Moorfield). Recovering the original names/terms of the Indigenous people, Wendt summons an important tie ('a fortified pa') to the islanders' ancestry, in which humans, islands, and plants are woven into mutual interconnectivity, 'to participate, act together, act in concert, join in (an undertaking)'. Earlier, in the face of a heating sun, a heliographic symbol of the nuclear radiation, the narrator tried to imagine 'how pre-otherworlder Tangata Maori would have seen the hill, the bush, everything' (189). Tangata is a Maori word for people. The word 'pre-otherworld' sets up a contrast against the totalitarian governance of a modern and colonial rule. The pre-colonial, pre-modern Maori people have not 'seen themselves as being separate from their surrounds,' not 'as the centre of it, selves who had the right to conquer, harness, subdue everything for their own use' (189). Empowered by the agaga of the Maori ancestors, the narrator's imagination and sense become 'keener than at any other time in [his] life' as he pronounces, '[e]very tree, vine, fern, stone, insect and bird, colour, smell and the everythingness that I felt I was becoming part of them' (189).

Wendt has had a poster of an atomic-bomb blast on his wall since 1965, and is a long-time friend of Maori poet Hone Tuwhare, whose early work was actively opposed to nuclear testing (Sharrad 207; 224 n.7). Not only does Wendt align himself with other Indigenous artists/writers of the Pacific, but retrieves the 'agaga,' the spirit of Maori ancestry and the Indigenous Pacific cultures, by re-inscribing the 'black rainbow' as a distinct cultural symbol of the Indigenous Pacific. Indeed, the concept of the 'black rainbow' demonstrates a paradox of representing nuclear radioactive abuse and Indigenous resistant energy. Wendt emphasises the rainbow's definition as an icon of Pacific diversity in that he foregrounds it as a figure for the mediation between the material and the spiritual worlds. In the 'black rainbow,' both as an icon of light and radiation and as a symbol of Indigenous ancestry, Wendt weaves 'flesh and history into sinews of the light that bound everything with its unbreakable genealogy. All creatures, all things, all elements' (*Black Rainbow* 242). As if performing a redemptive ritual, Wendt puts forth a similar gesture to Silko's, to tell a common story of nuclear plight, Indigenous solidarity and survivance. The narrator finally kneels to 'smell[ed] of dank sap, the mana which kept [him] alive' (*Black Rainbow* 243). Merged with his homeland and empowered by 'mana,' the supernatural force in the earth, he places the painting of Black Rainbow face up in a hole he digs. The Black Rainbow merges with the land, as the nuclear blackness is transformed into fertility and richness of the Indigenous earth. He reads the Hotere clock once more, then, as it ticked vigorously and pushes the earth onto it and articulates 'Whenua' (243). 'Whenua' in the Maori language refers to country, nation, territory, or domain. The Indigenous subject position emerges in the midst of the land and the people's resistance against colonial plight, from a unique perspective of 'tangata whenua,' i.e., people from the land or Indigenous people; 'tangata whenua' is to be natural, to be at home (Moorfield n. pag.). *Black Rainbow* employs Pacific discourses – Indigenous and re-invented – 'to create a postmodernism of resistance, revealing the power of performance' (Ellis 102). Wendt relies on methods of invention in Polynesian storytelling, which he identifies here as similar to the postmodern: 'All I know is, postmodernism isn't new if you look at the oral literature of Polynesia, wherein the teller and her personality become the story, and so forth' ('Pacific Maps' 61). As the narrator reads the Hotere clock again, the counting down is released from a definitive closure to become 'a counting backward to a starting point,' and this 'doubleness of the end meeting the beginning is reflected in the final chapter' (Sharrad 207).

Black Rainbow reflects a combined Pacific culture. The distinction between Maori and Pacific Islanders has been blurred as they are all Tangata Moni, 'merged with our sisters and brothers from the Islands ... [to become] the Tangata Moni, the True People' (158). 'Tangata' is the Maori word for 'people' and 'moni,' a Samoan phrase meaning 'true' (Moorfield). The combined Maori–Samoan naming suggests that Indigenous groups of the trans-Pacific claim the land of

Aotearoa/New Zealand as home (Bush 207; quoted in Alessioa 264). They are 'the original people' – 'they'll always be here' whether or not there is a 'monstrous' Memorial (12), exemplifying trans-Pacific survivance to act together to resist the threats imposed by global nuclear colonialism.

Pongso no Tau

Tau writer Syaman Rapongan's home island, *Lanyu* (Orchid Island) or in the Tau language, *Pongso no Tau*, is the site of anti-nuclear protests in the 1980s and a long-time part of Austronesian culture which spread across the Pacific.[7] Like Tangata Moni, Tau means 'the true people,' original inhabitants or first people of the island (Pangso). The nuclear waste storage facility in *Pongso no Tau* was built in 1982 and became the site to receive nuclear waste from Taiwan's three nuclear power plants operated by state Taiwan Power Company (Taipower). Syaman Rapongan participated in the anti-nuclear movement, first emerging in the late 1980s, when Indigenous rights activists protested the shipment of radioactive waste from nuclear power plants on Taiwan to *Pongso no Tau* (Cho 5–6). Like any small island of the Pacific, *Pongso no Tau* has been regarded as uninhabited, empty space, a closed ecosystem suited for nuclear activities.

Today, Syaman Rapongan articulates his powerful resistance in multiple mediums, including Facebook. For example, in a status-update written on 6 September 2015, he contrasts two photographs to lay bare nuclear colonialism on *Pongso no Tau*: (Figures 11.1 and 11.2).[8]

Figure 11.1 Status-update posted by Syaman Rapongan to his Facebook feed on 6 September 2015.

Figure 11.2 Status-update posted by Syaman Rapongan to his Facebook feed on 6 September 2015.

In his post, Rapongan states:

> The color both images seem the same blue and yet the one at the top [Figure 11.1] represents how the ocean embraces the island and creates a lively natural surrounding for the islanders whereas the bottom one [Figure 11.2] generates a representation which mimics the ocean by painting the cement wall blue and white. Within the wall is hidden the human-made radionuclides for nuclear waste. One is the living ocean the islanders cherish and the other, a misrepresentation invented by the 'international cheating group'.

Rapongan refers to the Taiwanese government (in collaboration with global corporations) as 'international cheating group'. In the same post, he adds that the nuclear waste-dumping site on *Pongso no Tau* is just like 'the human-made Bikini Atoll'. Rapongan calls attention to the destruction of *Pongso no Tau* and the challenges and struggles facing the Tau to the world stage by likening the nuclear dumping site to Bikini Atoll, where the U.S. detonated nuclear devices at test sites located inside the atoll, in the air, on the reef, or even underwater. He stands tall against the unchecked development and the greed of international corporations. In fact, the Taiwan Atomic Energy Council (AEC) has detected greater-than-class-C (GTCC) nuclear waste at the nuclear-waste storage facility on *Pongso no Tau*, despite the facility being designed for low-radioactive materials only, raising questions over the management of nuclear waste.

Through these two contrastive images, Syaman Rapongan conveys a message similar to Wendt's *Black Rainbow* and a profound sense that the Tau carry more than their share of the community's social burdens, and that something must be done. The dark clouds in Figure 11.2 speak of the internal darkness of grieving, a result of the Tau's nuclear plight. Rapongan showcases the daily life of the islanders to illuminate larger historical radiation ecologies initiated in the testing in the desert of the U.S. Southwest and the bombing of Hiroshima. His anti-nuclear message captures the public's imagination and, more importantly, the momentum of anti-nuclear activism in Taiwan. *Pongso no Tau* provides focus to anti-nuclear struggles not just in Taiwan but in the Pacific at large, signifying Taiwan's entry into the trans-Pacific activist politics.

Rapongan's memoir *Drifting Dreams on the Ocean*, composed of four chapters – 'Hunger of Childhood,' 'Cruising the South Pacific,' 'Voyaging the Moluques Strait' and 'Quest for Codes of Islanders,' – constitutes a navigator's chart of reflexivity and agency, criss-crossing the Pacific to seek partnership in his anti-nuclear activism. Rapongan sails out from his home island of *Pongso no Tau* to collect similar island stories en route, leading to a process of cultural awakening and recreation, a reclaiming of the past, its values and perspectives. Unique blending of cultural expressions and political action change the discourse of the islanders' rights and, in the process, pave the way for relevant issues to emerge. For instance, as he records in his travel logs, his encounter with a group of men in the Caroline Island in Micronesia brings him back to his traditional navigation knowledge which is also trans-Pacific in nature. The islanders take navigation as their occupation and demonstrate refined powers and extensive knowledge of astronomy, of the moon and the tides, with a degree of sophistication that rivals modern scientists and mariners: there are, for instance, 28 names for different kinds of winds one might encounter in the course of navigating the Pacific and numerous names for the currents and fishes: all this traditional knowledge maps a cosmic web connecting the water/underwater and the sky intimately while the human and non-human are conflated in the islanders' cosmology (194). Rapongan proclaims 'this is the environmental knowledge that the oceanic peoples share with the world' (194).

Rapongan makes explicit that it is an injustice against the Austronesian peoples in general, and Tau people specifically, keepers of this environmental knowledge, to be burdened with the waste generated by the nuclear plants. This distributive inequity also shows the toxic ethnocentrism of the Han (Chinese) people, who have 'turned the island into a colony of politicians and modern technology' (191). He desperately seeks the trans-Pacific community's support as many Pacific islands are facing common environmental justice challenges inflicted by nuclear colonialism of small islands. The memoir records several of his journeys across the Pacific in search of Indigenous solidarity, navigating the Pacific from *Pongso no Tau* to Tahiti among the Society Islands and other islands such as American Samoa, Samoa, Marshall Island, Solomon Island, and Fiji. His encounters with the Pacific islanders lead to an extraordinary expansion of Austronesian power and presence. As he conveys Tau values and messages through his stories, other islanders offer their testimonies:

> In 1996, many Polynesians on South Pacific islands went to the French government's administration building in Tahiti and protested against French government's secret nuclear testing in the low-island atolls, which goes against the international agreements. Mr. A participated in that campaign and later joined the German anti-nuclear association aiming to protect the planet as well as other campaigns that oppose the export of nuclear weapons to the Third World. In that year, the South-east wind from the South Pacific brought the radioactive fall-out of the testing to Society Islands, Cook Islands, and especially Rarotonga, where the capital city lies and where there are around 5000 inhabitants.
>
> (Rapongan, *Drifting Dreams* 195)

Like Pacific writer Epeli Hau'ofa, Rapongan envisions the Pacific/Oceania as 'a sea of islands,' protesting with voices rising from the nuclear seascapes. Both Indigenous and Oceanic bodies bear witness to the parallel exploitation of the ocean and its people. In the memoir, he recounts a walk he takes on Rarotonga. He comes across a local person, scaling some fish. He knew the name of the fish by heart as they share a common ocean: in Tau it is called *Mazowzaw*, and in Taiwanese it is called Goldband goatfish. He asks:

> 'Are these edible?'
> 'I don't know, but we've been wishing to eat fish for many years.'
> 'It's over ten years that we haven't eaten fish.'
> 'Is it safe to eat them?'.
>
> (Rapongan, *Drifting Dreams* 196)

The island Rapongan visited has been a site of nuclear detonations and toxic waste disposal. Nuclear-powered ships and vessels carrying radioactive materials ply the seas, while international businesses seek islands for the disposal of industrial wastes.

Like the Cook Islands that still struggle under the threat of radiation and the radioactive fallout suffered by Rarotonga, *Pongso no Tau* must endure a radioactivity that is polluting the sea and poisoning the ocean resources. Rapongan raises questions about this that are, of course, rhetorical:

> I really don't know how much the nuclear elements like plutonium, cesium, uranium, deathium will damage human bodies [...]. I actually feel the same way as that Polynesian did but could we just eat without giving it too much thought? For island people, not being able to eat fish is to cut their throats and force them to die.

Rapongan articulates a strong protest against increasing toxicity of Indigenous and Oceanic bodies which causes and constitutes genocide. The Pacific provides a significant proportion of the world's protein requirement and global reserves of minerals. The ocean is crucial for maintaining healthy human bodies. Yet, Rapongan's journey proves that pollution is causing just the opposite – illness.

The Tau calendar divides the year into three seasons, determined by the activities and life circles of the flying fish: *rayon*, the flying fish season (from March to early July), the time when men go out to the sea to catch the flying fish; *teiteika*, the end of the flying fish season; and *amyan*, the winter season is when men wait for the flying fish to return. Most of the flying fish caught during its high season are smoked and preserved as food sources for the rest of the year. On his home island, after the Flying Fish Season, Rapongan and other villagers used to catch coral fish in the coastal area that is now the nuclear waste repository; they used to dive and to spear fish on a daily basis to provide fresh fish to the household. It has never occurred to him or his people that the most agreeable food 'to [their] stomach and taste buds' would become poisonous. They were never informed by the Taiwanese government of the nuclear radiation (196).

Mr A, an islander Rapongan meets in his journeys, loathes not only the First World countries, which own nuclear power plants and nuclear weapons, but those who hold the knowledge and know-hows of high technology, for they are the local agents of imperialism and global capitalism that come to dominate and manipulate the Third World peoples (196). The islands and coral atolls of the Pacific not only serve as the repositories of the high-intensity nuclear waste and the sites of nuclear testing but they ironically become the popular tourist spots: Pacific islands, such as Bora Bora in Tahiti, Yasawa in Fiji, Hawaii, Guam, Yap, American Samoa, fall prey to the tourism dominated by Western countries and mega corporations (196). Through his fellow islander Mr A, Rapongan demonstrates a strong intention to disclose imperialism as a 'worlding' process that attempts to disguise its own workings so as to naturalise and legitimate Western dominance (Wilson 121); Mr A testifies 'the injustice that hegemonic nations and plutocrats make a good fortune for themselves by plundering the islanders' (*Drifting Dreams* 196).

As a custodian of the ocean in the face of colonial violence, Rapongan reaches out to similar people elsewhere 'in the common task of protecting the seas for the general welfare of all living things,' to use Hau'ofa's words (55). His writing constitutes a 'counter-worlding' in the sense that he evokes powerful local voices and sensibility and translates the Pacific as a figure, spacious and interconnected, which counteracts colonial imposition of nuclear violence.

Coda

In *Drifting Dreams on the Ocean*, Syaman Rapongan provides a holistic perspective in which the human and non-human are seen in the totality of their relationships, and the Pacific islands and archipelagos become a collective body. People raised in this environment are at home with the sea. They develop skills for building boats and navigation, as well as the spirit to traverse the large gaps that separate island groups (Huang 9). Rapongan draws our attention to the interconnectedness beneath the waves and in the sky, hence describing a deeper geography and mythology. The nuclear activism on Taiwan was initiated by the Indigenous cultural awakening of *Pongso no Tau* and through his writing and journey, Rapongan evidences a history, relatively free from pollution, injustice, and violence, longer than Western civilisation. His poetics is an ecologically interconnected, planetary, and re-nativised counter-conversion to islands and other seascapes. His act of crossing – crossing in multiple levels, physically, spiritually, geographically, and culturally – enables the trans-Pacific Indigenous solidarity based on their common struggles and survivance under colonial violence and erasure. His crossing also functions as opening to 'a species of alterity' that Gayatri Chakravorty Spivak names as the planet. The planet is, as always and already, a catachresis for inscribing collective responsibility as right: as Spivak puts it, 'to be human is to be intended toward the other' (73). Rapongan demonstrates what it means to be a true 'human' (Tau).

Likewise, In *Black Rainbow*, Albert Wendt opens up 'hidden histories' of the Pacific islanders: 'We are what we remember' (178). Invoking art forms and stories created by South Pacific islanders, including the Tangata Maori (the Maori people) and the fictional Tangata Moni (encompassing many Polynesian peoples and which Wendt 'translates' as 'the true ones'), his poetics is trans-Pacific and trans-Indigenous: 'My story, a collage history, contained in the ever-moving present' (189). By reclaiming 'the present,' 'ever moving,' for the Pacific peoples, Wendt re-situates the Indigenous subject in the present to write back on the colonial/modernist erasure of Indigeneity as backward and vanishing. He questions topographical and ideological constructs of the empire that attempt to reify culture and erase difference. In so doing, he effects a movement towards reclaiming a sense of connection to the islands and the Pacific first inhabitants.

Both Wendt and Rapongan underscore how the Indigenous people bear testimony to radiation ecologies and nuclear colonialism in the Pacific and imagine the Pacific connections in terms of the impacts of nuclear colonialism and militarisation on their home islands and the larger Pacific. Both focus on the survivance of 'the Indigenous Pacific' as a contact zone to examine the shifting relationships between empire and resistance. The islands and ocean will not go away. As Wendt puts it, the Tangata Maori have refused to become history ('Pacific Maps and Fiction(s)' 203). The Indigenous struggle for survival and self-determination is the great story of Pongso no Tau and Aoearoa. It is 'one of the great stories of the Pacific' ('Pacific Maps and Fiction(s)' 207).

Notes

1 Our co-authorship in this chapter took the following shape: Hsinya Huang translated the selected Sinophone texts of Syaman Rapongan's writing and Facebook posts and wrote the English script of this chapter; Syaman Ropongan contributed his tribal knowledge of the Tau people and his personal experiences with anti-nuclear activism in Taiwan, which Hsinya Huang rendered into English.

2 Some of this research can be found in *Aspects of Transnational and Indigenous Cultures*, co-edited by Hsinya Huang and Shu-Chung Chang. Also, in 2015, in collaboration with existing Humanities for the Environment (HfE) projects, this decade of collaboration laid the foundation for the organisation of the Asia-Pacific Observatory launched and funded by the Center for Humanities Innovation and Social Practices (CHISP) of National Sun Yat-sen University and the Humanities Center of National Chung Hsing University, based in Taiwan. See http://hfe-observatories.org/observatories/asia-pacific-observatory/

3 For a comprehensive timeline of nuclear weapons and denotations, see 'Nuclear Weapons Timeline', *International Campaign to Abolish Nuclear Weapons* (www.icanw.org/the-facts/the-%20nuclear-age/).

4 See Hau'ofa. Other references include van Fossen (*South Pacific Futures*); Robie (*Blood on Their Banner*); Fischer (*Stopping the Spread of Nuclear Weapons*); qtd. in Huggan and Tiffin 54–55.

5 Other Pacific literature written in English, which elaborate on related issues surrounding nuclear colonialism and militarisation in the Pacific, include Robert Barclay's *Meļaļ: A Novel of the Pacific*, Zohl de Ishtar's *Daughters of the Pacific*, James George's *Ocean Roads*, Chantal T. Spitz's *Island of Shattered Dreams* (Tahiti), Hone Tuwhare's *No Ordinary Sun* (Maori) as well as Kathy Jetnil-Kijiner's (Marshall Islands) and Craig Santos Perez's (Hawai'i) poetry, to name just a few.

6 Originally from 'Albert Wendt', 106; qtd. in Ellis 110 and Sharrad 206–207.

7 Indigenous Taiwan's significance in the world trans-Indigenous cultures cannot be over-emphasised though contemporary scholarship in Indigenous studies seems to overlook the contributions from this part of the world. The Austronesian world, originated in Taiwan, spreads across the Pacific and Indian Ocean, north to Taiwan, south to New Zealand, east to Easter Island and west to Madagascar. The Polynesian people are considered to be by linguistic, archaeological and genetic ancestry a subset of the sea-migrating Austronesian people. As S. R. Fischer notes, Micronesians and Polynesians owe most of their genetic and cultural make-up to Austronesian ancestors (11–12). Recent scholarship has identified an 'out of Taiwan' hypothesis of Austronesian expansion by tracing the travels of paper mulberry seeds (see Matisoo-Smith). Others argue that Taiwan has ushered in the Austronesian languages that became about 1,200 in number spreading across most of Oceania, the Pacific, and the Indian Ocean (see Diamond). Still others forge the close connections between seafaring tradition and Austronesian origins of Oceania (see Rolett et al.).

8 These pictures can be found in colour on Rapongan's Facebook page, see https://goo.gl/YV6LJS.

Works cited

Alessioa, Dominic. 'From Body Snatchers to Mind Snatchers: Indigenous Science Fiction, Postcolonialism, and Aotearoa/New Zealand History', *Journal of Postcolonial Writing* 47.3 (2011): 257–269.

Barclay, Robert. *Melal*. Honolulu: University of Hawai'i Press, 2002. Print.

Bush, Barbara. *Imperialism and Postcolonialism*. Harlow: Pearson Longman, 2006. Print.

Carson, Benjamin D. 'Ainu and Anishinaabe Stories of Survivance: Shigeru Kayano, Katsuichi Honda, and Gerald Vizenor'. 東アジア文化交渉研究 2 (2009): 443–449. Print.

Cho, Yu-Fang. 'Nuclear Diffusion: Notes Toward Reimagining Reproductive Justice in a Militarized Asia Pacific', *Amerasia Journal* 41:3 (2015): 2–24.

Dé Ishtar, Zohl. *Daughters of the Pacific*. North Melbourne: Spinifex, 1994. Print.

DeLoughrey, Elizabeth M. 'Heliotropes: Solar Ecologies and Pacific Radiations', *Postcolonial Ecologies: Literatures of the Environment*. Eds Elizabeth DeLoughrey and George B. Handley. Cambridge: Oxford University Press, 2011. 235–253. Print.

——'The Myth of Isolates: Ecosystem Ecologies in the Nuclear Pacific', *Cultural Geographies* 20 (2013): 167–184. Print.

——'Radiation Ecologies and the Wars of Light', *Modern Fiction Studies* 55.3 (2009) 468–495. Print.

Diamond, Jared M. 'Linguistics: Taiwan's Gift to the World', *Nature* 403 (2000): 709–710.

Ellis, Juniper. 'Postmodernism of Resistance: Albert Wendt's *Black Rainbow*', *ARIEL* 25.4 (1994): 101–114. Print.

Fischer, David. *Stopping the Spread of Nuclear Weapons: The Past and the Prospects*. London: Routledge, 1992. Print.

Fischer, Steven Roger. *A History of the Pacific Islands*. New York: Palgrave, 2002. Print.

George, James. *Ocean Roads*. Wellington: Huia, 2006. Print.

Hau'ofa, Epeli. *We Are the Ocean: Selected Works*. Honolulu: University of Hawaii Press, 2008. Print.

Huang, Hsinya. 'Representing Indigenous Bodies in Epeli Hau'ofa and Syaman Rapongan', *Tamkang Review* 40.2 (2010): 3–19. Print.

Hsinya Huang and Shu-chun Chang, eds. *Aspects of Transnational and Indigenous Cultures*. Newcastle upon Tyne: Cambridge Scholars Publishing, 2014. Print.

Huggan, Graham and Helen Tiffin. *Postcolonial Ecocriticism: Literature, Animals, Environment*. New York: Routledge, 2010. Print.

Ibuse, Masuji. *Black Rain: A Novel* (黒い雨). 1966. Trans. John Bester. Tokyo: Kodansha, 1969. Print.

Matisoo-Smith, Elizabeth A. 'Tracking Austronesian Expansion into the Pacific via the Paper Mulberry Plant', *PNAS* 112.44 (2015): 13432–13433. Print.

Moorfield, John C. *Te Aka Online Māori Dictionary*. N.d. Web.

'Nuclear Weapons Timeline', *International Campaign to Abolish Nuclear Weapons*. N.d. Web.

Rapongan, Syaman. 大海浮夢 [*The Drifting Dreams on the Ocean*]. Taipei: Union Publisher, 2014.

——Facebook Post. N.d. Web.

Robie, David. *Blood on their Banner: Nationalist Struggles in the South Pacific*. London: Zed Books, 1989. Print.

Rollett, Barry V. 'Taiwan, Neolithic Seafaring and Austronesian Origins', *The Free Library*. 2000 Antiquity Publications, Ltd. Web.

Sharrad, Paul. *Albert Wendt and Pacific Literature: Circle the Void*. Manchester: Manchester UP, 2003. Print.

Silko, Leslie Marmon. *Ceremony*. New York: Penguin Books, 1977. Print.

Spitz, Chantal T. and Jean Anderson. *Island of Shattered Dreams*. Wellington: Huia, 2007. Print.

Spivak, Gayatri Chakravorty. Death of a Discipline. New York: Columbia University Press, 2003.

Tuwhare, Hone. *No Ordinary Sun*. Dunedin: McIndoe, 1977. Print.

van Fossen, Anthony B. *South Pacific Futures: Oceania Toward 2050*. Brisbane: Foundation for Development Cooperation, 2005. Print.

Vizenor, Gerald. *Fugitive Poses: Native American Indian Scenes of Absence and Presence*. Lincoln: University of Nebraska Press, 1998. Print.

——*Hiroshima Bugi: Atomu* 57. Lincoln: University of Nebraska Press, 2003. Print.

——*Manifest Manners: Narratives on Postindian Survivance Earthdivers: Tribal Narratives on Mixed Descent*. Minneapolis: University of Minnesota Press, 1981. Print.

——*Survivance: Narratives of Native Presence*. Lincoln and London: University of Nebraska Press, 2008. Print.

Wendt, Albert. 'Albert Wendt'. *In the Same Room*. Eds Elizabeth Alley and Mark Williams. Auckland: Auckland University Press, 1992. 100–118. Print.

——'Pacific Maps and Fiction(s): A Personal Journey', *Perceiving Other Worlds*. Ed. Edwin Thumboo. Singapore: Unipress, 1991. Print.

——*Black Rainbow*. Honolulu: University of Hawaii Press, 1992. Print.

Wilson, Rob. *Be Always Converting, Be Always Converted: An American Poetics*. Cambridge, Mass.: Harvard University Press, 2009. Print.

12 Walking together into knowledge

Aboriginal/European collaborative environmental encounters in Australia's north-east, 1847–1849

Michael Davis

Introduction

In the mid-nineteenth century naturalists and botanists scoured the colonial world securing 'natural history' specimens to satisfy the imperial science project's insatiable thirst for classifying and collecting. One contributor to this was the British ship HMS *Rattlesnake* which, during 1847 to 1849, led an expedition to survey Australia's tropical north-east coasts, islands and reefs, and to assess the suitability of the newly encountered land for British settlement and expansion.

In this chapter I examine some episodes in the collaborative engagement between the voyagers and local Aboriginal people during the *Rattlesnake* expedition, with a specific focus on Aboriginal peoples' plant and animal knowledge. I juxtapose these nineteenth century colonial collaborative encounters with a brief look at the gathering together of many nations in Paris in December 2015 that resulted in a collaborative agreement to counter the global impacts of climate change, to argue that both of these scenarios illustrate the kind of integration of knowledges and rethinking of human/environment relationships that is called for in Anthropocene thinking within the environmental humanities.

The voyage of the *Rattlesnake* is well documented. Jordan Goodman's 2005 book *The Rattlesnake: A Voyage of Discovery to the Coral Sea* provides a detailed narrative of the expedition, beginning with its departure from Plymouth, England in late 1846. Iain McCalman's *The Reef: A Passionate History* presents a nuanced discussion (at 117–140) of the complex encounters between Aborigines, *Rattlesnake* voyagers and a young woman called Barbara Thompson, who had been shipwrecked and was living among the local people. The *Rattlesnake* voyage produced a large volume of archival records including diaries, journals, shipping records and correspondence, much of which documents the voyagers' observations of and encounters with the local Indigenous people.[1] Local Aboriginal people were, at times in the expedition records, described as collaborators, or companions, as they accompanied the expeditioners on their journeys through the country in search of new species, and their deep knowledge of the environment was useful for the expedition's botanical and natural history collecting.[2]

In examining encounters between British explorers and local Aboriginal people, I propose these as instances of what I call 'walking together,' or moments in which humans of diverse cultural backgrounds met with one another to engage with other humans and nonhuman species in the common purpose of knowing and interpreting the local environment.[3] Although the *Rattlesnake* expedition's natural history project was primarily to *collect* the species that were found, my focus here is on the ways that these explorers described their engagement with these species, and with local Aboriginal knowledge in their encounters with Aboriginal people. I suggest that these encounters can be interpreted in the context of ideas about multispecies ethnography (see for example Faier and Rofel; Ogden, Hall and Tanita; Kirksey and Helmreich), and the ideas of leading environmental humanities scholars Deborah Bird Rose and Val Plumwood. Analysis of the early nineteenth century collaborations between local Aboriginal people and colonial British explorers also will reveal some of the roots of the globally linked Humanities for the Environment (HfE) Observatories in philosophical concepts of 'Country' (see Rose, Chapter 3 this volume). I frame this history as pointing towards what we are calling in this volume 'new constellations of practice' (cf Adamson, Introduction), that forge different ways of working together in trans-disciplinary formations.[4]

In what follows, I draw an analogy between these mid-nineteenth century collaborative encounters between Aboriginal people and explorers in north-east Australia, and what I call a 'walking together' moment that took place when many nations met in Paris in December 2015 to agree to a plan to reduce the global impacts of human-induced climate change. The Paris meeting, though not construed as a 'multispecies encounter,' can be envisaged, as with the *Rattlesnake* Aboriginal/explorer meetings, as an instance in which diverse actors 'walked together' in environmental conversations and negotiations. My purpose will be to explore what these two widely different scenarios illustrate about how multiple, diverse actors can coalesce in moments of common environmental purpose, whether it is to reach agreement about addressing climate change, or to engage over the identification of new species. Recognising these kinds of connections between and among the diversity of humans, and other, nonhuman species, and learning about new kinds of knowledge, is a significant contribution that the humanities and social sciences can make to the sciences. It can also be thought about in the context of the Common Threads (which include 'The Role of Humans' and 'Humanities Innovation') that connect the Humanities for the Environment (HfE) Observatories.[5] The kinds of human/nonhuman environmental coalescing in the Aboriginal peoples' encounters with nineteenth century British naturalists and explorers that I illustrate in this chapter can be understood within these Common Threads. A Common Thread that connects the *Rattlesnake* environmental encounters to the Paris meeting is the role of Indigenous knowledge. Indigenous knowledge had a central role in the meetings between *Rattlesnake* naturalists and local Aboriginal people, as they walked together in the Australian 'bush'[6] in the pursuit of plants and animals. In the Paris Agreement, Indigenous knowledge has a central role in some of the provisions on mitigating the effects of climate change, such as those on biodiversity protection and adaptation.

I want to ask in this chapter, what might these two collaborative encounters, separated in time and place, from nineteenth century tropical Queensland to twenty-first century Paris, suggest about a rethinking of 'what it means to be human, and humans' relations with the natural world?' And what do they imply in the context of environmental humanities shaping of 'new constellations of practice' around intersecting knowledges? Val Plumwood argues that the spiritual and ethical failures of modern society, and resulting historical separation of humans from nature, has produced a kind of 'ecological denial'. She articulates an alternative approach for an 'environmental culture' as a decentering critique of the hegemony of mono-cultural, anthropocentric models, one that reconnects humans with ecology. In Plumwood's thesis, the kind of deeper belonging articulated by Aboriginal people as central to their relationships with land, that incorporates a communicative and spiritual dimension, is also critical to developing her critique against monological, hierarchical systems and human–nature separation. As she states 'the communicativity and intentionality of more-than-human others is often the key to the power of place' (230). Both Plumwood, and Deborah Bird Rose, as we see in Rose's chapter in this volume, are particularly attentive to Aboriginal concepts and ways of belonging in the land, ideas crucial to my discussion of Aboriginal-explorer encounters in nineteenth century Australia. Rose has articulated the idea of 'Country' in the following way: Country is multi-dimensional: it consists of people, animals, plants, Dreamings, the dead, and the yet to be born, underground, earth, soils, minerals and waters, surface water, and air (*Reports from a Wild Country* 153).

This notion of 'Country' as something more than 'landscape' or 'territory,' which is multi-faceted, and deeply embedded within Australian Aboriginal culture, forms the basis for the connections that Aboriginal people have to place and landscape, and is important to bear in mind when examining the environmental encounters between local Aboriginal people and *Rattlesnake* expeditioners. The Aboriginal idea of 'Country' as articulated by Rose also echoes the ways of 'knowing the earth' sustained by Chicaza people as discussed by Linda Hogan (Chapter 2, this volume). It is a long way from nineteenth century colonial Australia to Paris in 2015; but these two scenarios share some themes which point to my argument about 'walking together' in environmental conversations.

Walking together: 1840s Australia and 2015 Paris

The journey of HMS *Rattlesnake* through north-east Australia took place at a time when there had already been some European presence in the region, but just before the era of more intensive settlement and frontier violence towards Aborigines.[7] The expedition followed earlier voyages of survey and discovery through this region by James Cook (1770), Matthew Flinders (1799, 1802), Phillip Parker King (1819–1820) and others, seeking sites for shipping, trading, watering and refuelling stations; and for ports and harbours that could serve the coming European pastoral and settlement expansion. The naturalists on the *Rattlesnake* voyage sought to locate, identify and classify and to collect 'specimens,' especially those they thought to be 'new to science'. In one sense this collecting advanced the British Empire's exploitative, extractive project, with the concomitant effects of obscuring or effacing Indigenous peoples' autonomy, their connections to 'country,' their knowledges, and their agency and presence. In this way of interpreting the records, although sometimes portrayed in voyaging records as 'collaborators,' or 'companions,' local Aboriginal people were in fact disingenuously conscripted to the imperial cause, passively compliant in the erasure of their own presence in the tropical environs. But I offer a reading of the *Rattlesnake* journals in which these encounters can instead be considered as brief moments of mostly consensus-based collaboration between local Aboriginal people and the British, with a common purpose in engaging around the 'knowing' of species, their properties and habitats. Acknowledging that natural history has erased the contributions of Indigenous knowledge keepers, I will read the *Rattlesnake*'s records for what they show us, instead, about collaboration or 'walking together'.

Another, very different kind of 'walking together' took place in Paris in December 2015. After two long weeks of intense negotiations, approximately 190 nations finalised an international Agreement to slow the rapidly increasing impacts of human-induced climate change. The Paris Agreement (formally known as the 21st session of the Conference of Parties to the United Nations Framework Convention on Climate Change) provides, among other things, for developing and developed countries to limit global warming to no more than 1.5 degrees. The Agreement includes language on finding a rapprochement between Indigenous knowledge and 'Western' science in its opening paragraph. This notion of 'integrating knowledges' is expressed in the Agreement's provision (Article 7(5)) regarding adaptation, which calls for a:

> country driven, gender-responsive, participatory and fully transparent approach, taking into consideration vulnerable groups, communities and ecosystems, and should be based on and guided by the best available science and, as appropriate, traditional knowledge, knowledge of indigenous peoples and local knowledge systems, with a view to integrating adaptation into relevant socioeconomic and environmental policies and actions, where appropriate ('Adoption of the Paris Agreement').

This paragraph reflects the kind of thinking that responses to the Anthropocene might encourage, that is, to address human impacted climate change globally we need to reassess the nature of relationships between disparate ways of knowing, and between humans, nonhumans and other objects. The opening paragraphs to the Paris Agreement also ask Parties (countries) to 'note the importance of all ecosystems, including oceans, and the protection of biodiversity, recognized by some cultures as Mother Earth …'. These aspirational words call for an engagement between the kinds of technical-rational, scientific approaches to biodiversity protection, and those that are more Earth-centred, based on Indigenous ways of perceiving the natural world. Yet despite the language of recognition at this global level of United Nations discourse, as Tony Birch (Chapter 13, this volume) points out, there is still a relative absence of proper recognition for Indigenous peoples and their knowledge systems in the growing scholarly debates and discussions around climate change. The Paris Agreement's call for a working together, or integration, of scientific and Indigenous knowledges recognises the centrality of such an integration to address the global impacts of climate change. There is also an underlying suggestion in the wording of these parts of the Agreement, that by asking nations to be 'guided by … traditional knowledge, knowledge of indigenous peoples and local knowledge systems,' they might be encouraged to bring a sense of wonder, or at least curiosity into their working together of scientific and Indigenous knowledge systems. This integration of knowledges can be illustrated by looking more closely at some specific examples of encounters between local Aboriginal people and British explorers in nineteenth century Australia.

Explorers, local Aboriginal people and multispecies ethnography

One Thursday at the start of November 1849, Charles Card, a young English born midshipman and ship's clerk with the *Rattlesnake*, was impressed by his encounter with Aboriginal knowledge. In his diary for Evans Bay in Queensland's Cape York Peninsula, he recorded that he 'took a stroll into the lagoon accompanied by two or three natives and among them [one] who called himself 'Tchow' [?] and one who was a great chum of mine'. Here, Tchow 'got me some kind of red fruit with a large stone in the middle which he made signs for me to eat, which I did and found very good'. The 'old man,' said Card, 'was quite a botanist having a different name for every different kind of tree and seemed very proud of his knowledge'.[8] Card was also shown local Aboriginal methods of procuring turtle,[9] and his companion Tchow also took him to a place where there were 'plenty of green parrots'.[10] From these descriptions, Card's diary shows him to be gradually beginning to immerse himself in the expert knowledge that Tchow readily displays about the local environment.

As the *Rattlesnake* expedition continued on its long voyage northwards through the tropics, Charles Card and the ship's naturalist John MacGillivray became increasingly curious about the local Aboriginal peoples' environmental practices and knowledge. At Evans Bay on Cape York Peninsula, where the ships stayed for some time in late 1849, Tchow showed Card how to obtain some native honey. Card reported that he and others from the ship 'went on shore for a good long day's cruise' and hunted for birds. Each man 'took a native,' as they phrased it, with 'old Tchow being my *kotiega* [Card's emphasis]'. MacGillivary translated this term (referring to it as *kotaiga*; also *Cotaiga*), suggesting that it derived from a Kuruareg (one of the Cape York language groups) word 'kutaig' for younger brother, but was used by the ship's personnel to mean 'friend,' 'associate' or 'companion' (MacGillivray I 310). As the party were returning to camp, Card's companion Tchow 'suddenly stopped and pointed to the dead trunk of a tree which was laying on the ground'. Card relates the event:

> I could see nothing particular about it until he pointed to a small hole from which several small flies were going in and out; as he proceeded to knock a larger hole I went some distance thinking I should be stung, but the old man soon called me back and I sat down and watched him; he first of all got a large stick with which he made a hole in the tree large enough to put his arm in, and having done so he hauled out a handful of dark looking honeycomb from which the honey was dropping fast; it was also covered with eggs and old and young flies, all of which went into his mouth together and very soon disappeared; he then got me a piece, clear of eggs &c and which I found very good but not so good as English honey ...[11]

This scene shows multiple human to human and human–environmental connections: between diverse cultural backgrounds and between peoples and nonhuman species. In a multispecies ethnographic interpretation, in this brief moment, Card and Tchow are connected in time and place to the nonhuman species that includes the dead tree, the flies, the honeycomb and eggs. There is a sensory dimension to this encounter that gives rise to Card's intimation of wonder and curiosity, and, which informs Tchow's demonstration of his deep knowledge of the local environment and its capacity to provide food resources. Card is, at first, fascinated by the apparently sudden appearance of the honey from the dead tree, by Tchow's adept know-how in finding this cache of native honey and extracting it, and by the Aboriginal man's eager consumption of the raw honey covered with eggs and flies. Concluding his observation, having himself by then tasted the honey, Card then reverts to a comparative frame of reference bounded by his familiar, Old World categories, as he reports that the native honey was 'not so good as' the English variety.

The Gudang man Tchow had become a companion to Card, a helper, and guide, and possibly interpreter to the young sailor. He seems to have been a leader among the people in the Evans Bay region, and was described by ship's artist Oswald Brierly, as a 'fine stout man, tall and strongly made, with a very good expression' (Moore 90). In Brierly's account Tchow (whom he records as 'Tchiako') was 'born on Moralug [Prince of Wales Island], the ground or place that he inherits from his father is Moreelug, (Adolphus Is.) but is the principal man of Evans Bay and neighbourhood' (Moore 91; McCalman 117–140). Evans Bay seemed to be 'a traditional meeting place for the Gudang' (Goodman 181), and visitors from other places including the nearby Torres Strait Islands and Papua. MacGillivray claimed in May 1849 that 'At Cape York we had as many as 200 natives (of various tribes) assembled at one time'.[12]

In the human/environmental engagement such as that between Card and Tchow and the native honey, Aboriginal people, British explorers and the myriad of species they are engaging with momentarily coalesce into what might be considered as 'multispecies assemblages' (Ogden et al.). We can read this mutual engagement around the native honey within a frame of 'multispecies ethnography,' as a way of interpreting the ways in which 'the host of organisms whose lives and deaths are linked to human social worlds' (Kirksey and Helmreich 545). In sites of colonial encounter such as described in Card's account, with their complex transactional social, economic and political contexts, there is a comingling of humans of diverse cultural backgrounds, together with multiple species. As Kirksey and Helmreich put it 'a multitude of organisms' livelihoods shape and are shaped by political, economic, and cultural forces' (545). Ethnographic readings of multispecies encounters such as these, in Ogden, Hall and Tanita's argument, have 'chipped away at the essentialism of nature' and 'focused on the ways in which naturalized environments reverberate with cultural significance, acting as repositories of cultural memory, false memories, mythology, social identity, and as sites of production and reproduction …' (11). Here, 'instead of solely conceptualizing human difference within an array of human categories, multispecies ethnographers are conceptualizing the human as a register of difference that emerges through shifting, often asymmetrical, relations with other agentive beings' (7).

The *Rattlesnake*'s natural history collecting advanced the Enlightenment project's purported goals of 'improvement' and 'progress,' as Richard Drayton has explicated these ideas in his book *Nature's Government*. European natural history collecting at the time of the *Rattlesnake* expedition was informed by a view of nature as outside humanity, and outside history, to be subdued and collected, useful for the economic advancement of the empire. Yet in brief moments of mid-nineteenth century encounter such as that between Charles Card and Tchow, 'the environment' with its multitudes of species, assumes a kind of agency for the British collector, as it becomes a source for wonderment and curiosity, as well as a site for collecting new species for societal improvement. It is these kinds of expressive, emotional representations in the voyaging narratives, as depicted in Card's diary, that support my argument that encounters such as these can be understood as multispecies moments. Brought into current discourses surrounding the Anthropocene and the environmental humanities, examining these encounters also allows us to 'extend the conversation,' as Hogan and Rose, and also Adamson suggest in their introductory chapters in this volume, about ways in which we can better understand Anthropocene-related activities that have contributed to increasing environmental change in the 'Age of the Human'. Within a framework of multispecies encounters and 'new constellations of practice,' the role of Indigenous knowledge in the meetings between British explorers and local Aboriginal people is interpreted in a deeper, multi-dimensional way, as it contributes to a complex that also includes place, and movement, a theme I will tease out some more below, and which is also explored in Chapter 4.

The kind of collaboration that the *Rattlesnake* voyagers developed with local Aboriginal people is not uncommon in the historical records. Aboriginal peoples' assistance was often invaluable to explorers, leading them to much needed water, and acting as guides and interpreters (see for example Clarke). In a comparative study of early exploration in Africa and Australia, historian Dane Kennedy writes that local knowledge 'proved essential to explorers' abilities to overcome the challenges they faced in the field' (2). He states that 'the abler explorers quickly learned that their prospects for success – indeed, their prospects for survival – depended in large measure on their ability to acquire access to local knowledge' (193–194). But the records of the *Rattlesnake* expeditioners Charles Card and John MacGillivray suggest that local Aboriginal knowledge was not only useful; it was also perhaps understood, or acknowledged, in other, more emotional registers by the Europeans, as will be seen in the example that follows. Writing to the British ornithologist John Gould in late February 1850, MacGillivray stated that 'it was chiefly by our intimacy with the natives that we succeeded in getting so many new birds'.[13] This notion of 'intimacy' may imply a deeper engagement with local knowledge. Similarly, Charles Card's encounter with Tchow suggests that curiosity and at times, wonder, also informed the intruders' responses to their encounters with Indigenous knowledge. Becoming 'intimate' with local Aboriginal knowledge facilitated the explorers' imperial project in identifying and collecting species for science. But at the same time, it also conveys a sense of wonder, as the British were becoming more closely engaged with this knowledge.

A deeper understanding of local Aboriginal knowledge is also illustrated in MacGillivray's account, where at one place he showed respect for Aboriginal peoples' taboo associated with a particular species. At Moreton Bay (today's Brisbane) he noted a certain 'undescribed porpoise, a specimen of which, however, I did not procure, as the natives believed the most direful consequences would ensue from the destruction of one'. He wrote 'I considered the advantages resulting to science from the addition of a new species of *Phocaena*, would not have justified me in outraging their strongly expressed superstitious feelings on the subject' (MacGillivray I 48). This animal, known by its common name as the Harbour Porpoise, is one of several species of porpoise widespread in the Pacific and other oceans. The closer engagement with local Aboriginal knowledge that MacGillvray came to acquire as the *Rattlesnake* expedition proceeded provided the grounds for his more cautious approach, and reluctance to secure specimens of this species. The naturalist here seemed able to steer a course between the demands of Western science for the procuring of specimens, and his appreciation of local Aboriginal cultural protocols. This combination of wonder, curiosity, and inquiry provided the essential ingredients for an integration of knowledges; an integration also articulated in the Paris Agreement, and which goes to the heart of Anthropocene thinking.

Collaborations of the kind described between the *Rattlesnake* voyagers and local Aboriginal people presented opportunities for the British explorers to observe and experience the natural world and to learn to appreciate the local peoples' ways of engaging with the environment. In the collaborative encounters, Card's companion Tchow was not only a source for local environmental know-how, he also assisted Card in other ways. As the ship's clerk described, he 'carried my haversack which contained everything except the gun'.[14] Brierly also refers to another *Cotaiga* or friend, a Gudang man called Billadi, (or Billiedi). Writing in October 1848, Brierly observed 'I had not been long on shore when my *cotaiga* of yesterday came running along the beach to meet me and took possession of my haversack and drawing material as a sort of matter of course' (Moore 32, emphasis in original). These *kotaiga/kotiiega*-type collaborations developed further on the *Rattlesnake*'s return to Evans Bay in October 1849. At this point, 'Immediately on landing for the purpose of an excursion,' wrote MacGillivray, 'each of us looked out for his *kotaiga*' (I 310). MacGillivray's *kotaiga*, who was called Paida, 'considered himself bound to attend to my safety, so conducted me to the boat which he assisted in shoving off, nor did he retire from the beach until we had got into deep water' (I 312). Paida accompanied MacGillvray on many of his walks into the surrounding country and, in a similar way to the relationship between Card and Tchow, became his guide and assistant in the botanising and identification of plants and animals.

In their collaborative engagement around plants and animals, Aboriginal people and British voyagers are also working together in the co-production of knowledge that intricately connects them with place (see for example Ingold *Lines; Being Alive*). The *Rattlesnake* encounters also implicate *place* in their presentation as fuzzy assemblages where multitudes of things coalesce. John MacGillivray is 'walking into knowledge' as he is journeying on foot through the newly experienced tropics, finding, classifying and collecting plants and animals, many of which he thought to be 'new to science'. MacGillivray is walking though the bush in tropical Australia, intent on adding to his collection. At the same time, he also experiences the country. Taking anthropologist Tim Ingold's notion of 'wayfaring,' we can imagine that, as MacGillivray walks though the hot countryside, he is alive to the experiences of what he observes, hears, and feels around him. At the same time, he is absorbing information, not only by way of traversing the landscape, but also instrumentally, through his walking and collecting. Ingold elaborates on the centrality of *movement* in conversations about place, arguing that 'lives are led not inside places but through, around, to and from them, from and to places elsewhere' (*Being Alive* 148). Wayfaring then, is 'the embodied experience of his perambulatory movement' (Ibid). The walker as wayfarer is becoming knowledgeable; he is inhabiting the earth and coming to know it. Ingold explains elsewhere that:

> For the walker, movement is not ancillary to knowing – not merely a means of getting from point to point in order to collect the raw data of sensation for subsequent modelling in the mind. Rather, moving *is* knowing. *The walker knows as he goes along.*
>
> (*Lines* 47, Ingold's emphasis)

This has important implications for knowledge production, again, following Ingold, who says that knowledge (both scientific and local) is 'integrated not through fitting local particulars into global abstractions, but in the movement from place to place, in wayfaring' (*Being Alive* 154). He expands on this:

> Rather than treating science and culture as equal and opposite, ranged on either side of an arbitrary division between space and place, and between reason and tradition, a better way forward … would be to acknowledge that scientific knowledge, as much as the knowledge of inhabitants, is generated within the practices of wayfaring.
>
> (*Being Alive* 155)

Henry David Thoreau, one of the great walkers and chronicler of woodlands, also understood the relationships between knowledge, sensibility, and place. He wrote 'I have met with but one or two persons in the course of my life who understand the art of Walking, that is, of taking walks … who had a genius, so to speak, for sauntering …' (251). Thoreau understood the way that being in the woods – a familiar place – restored, or brought about a connection to the inner being, when he claimed:

> I think that I cannot preserve my health and spirits, unless I spend four hours a day at least – and it is commonly more than that – sauntering through the woods and over the hills and fields, absolutely free from all worldly engagements.
>
> (254)

Thoreau also ruminates on the predicament of walking into the woods, but *not being in the woods*, in that particular place and moment, in a sensory, or phenomenological sense:

> I am alarmed when it happens that I have walked a mile into the woods bodily, without getting there in spirit. In my afternoon walk I would fain forget all my morning occupations and my obligations to society. But it sometimes happens that I cannot easily shake off the village.
>
> (258)

He claimed of the alarming proposition that 'the thought of some work will run in my head and I am not where my body is – I am out of my senses' (258–259; see also Adamson, Chapter 8, this volume).

Conclusion: walking together for the earth

Aboriginal people and British voyagers had vastly different motivations for entering into these kinds of collaborative engagements around the natural world. Aboriginal peoples' motivations for collaboration and friendship may have been, in part at least, possibly stimulated by their desire for trade goods, and possibly also to establish alliances with the visitors, in local political manoeuvres vis-à-vis other Aboriginal groups. As for the British, their collaboration was almost certainly driven by an economic imperative, in the context of the imperial vision for assessing the newly discovered lands; and they were also keen to engage with local people to assist the botanical-scientific collecting project. Notwithstanding these different motivations, the British explorers and the Aboriginal people's collaboration melded in encounters 'for the Earth's sake'. In this sense, these meetings present a possible scenario for thinking more widely about connections between peoples, and between humans and the environment. In these encounters in tropical Australia we see an emergent multispecies ethnography that connects diverse humans to other nonhuman species in contexts of negotiated intercultural transactions. They present a mutual engagement of the kind that might be paralleled with that witnessed at the Paris Climate Change meeting. There, in the heart of Europe, in December 2015, peoples from many diverse nations 'walked together' in common purpose, to reach agreement on ways to reduce the rapidly encroaching crisis of global human-induced climate change. In tropical north-eastern Australia, in October 1849, peoples from at least two very diverse nations – British voyagers, and several Aboriginal cultural/language groups – 'walked together' in the mutual endeavour of identifying, observing (and especially for the British, collecting), and maintaining practices relating to, plants and animals.

If historical actors such as the British naturalists and ships' clerks on the *Rattlesnake* were able to enter into complex, negotiated collaborations with local Aboriginal people in a mutual project to identify and name the natural world and its properties, then might 'walking together' become a device/model for the kind of re-integration between humans and nonhumans that is sought within the environmental humanities by Anthropocene thinking? If Anthropocene thinking is compelling us towards a re-orientation of human–nonhuman relationships, then this also presents the foundations for re-situating our thinking about all other kinds of relationships: cross-cultural, Aboriginal/European, colonial/ postcolonial. The observations and cross-cultural environmental knowledge encounters that I have examined here are played out in the microcosm of a mere few years in the mid-nineteenth century, in a setting bounded geographically within a relatively small area of north Queensland. These localised, historical-scale Aboriginal/British encounters and textualised observations have far reaching implications within the context of today's Anthropocene thinking. The kinds of collaborative relationships that were formed between the *Rattlesnake's* voyagers and individual Aboriginal people, in settings saturated with discoursing on environmental knowledge, offer a springboard for imagining new types of relationships between humans cross culturally today, and between humans and the nonhuman world.

At the Paris Climate Change meeting, multiple actors 'walked together' in conversations and negotiations about species, environments, environmental change, biodiversity, adaptation, and Indigenous knowledge. The Paris Climate Change summit brought together a wide range of actors in addition to the representatives of nation-states. This, argues David Cash in the on-line journal *The Conversation*, made the Paris Agreement a 'game-changer, aligning countries in a way that had not been accomplished before'. He writes that 'this included a traditional horizontal alignment across countries and also a novel call for an alignment from global to local levels' (Cash n.pag). What also characterised the Paris meeting, suggests Cash, was the involvement of multiple 'non-party' actors (non-government, including civil society, private sector, financial institutions, local communities and Indigenous peoples). He states that although they were not signatories to the Paris Agreement, 'they were fully embedded in the Paris Agreement itself' (Cash n.pag).

Can we envisage the nineteenth century Aboriginal/Voyager encounters over species, and the 2015 Paris Climate Change meeting, in their very different ways, as each hinting towards possible 'knowledge integrations,' local and global, that knit together diverse humans and species in what Ogden et al describe as Anthropocene inspired 'new forms of environmental governance and institutions' that might allow 'the development of novel approaches for more holistically engaging with a broader and necessarily more complicated articulation of Earth Stewardship' (341, 345). These re-connections between the human and the natural worlds would allow an Earth care, in terms articulated by Deborah Rose (this volume) in her description of the Aboriginal idea of Country, and by Linda Hogan (this volume) in terms of a First Nations Chicaza world view wherein 'the human and the earth body have the same cartography, movement and life energy' and 'the beauty of the world is the equal to the human place within it'.

Notes

1 In this Chapter I use the term Indigenous when referring generally to Australia's First Peoples, which includes Aboriginal people, and Torres Strait Islander people. I use the term Aboriginal when discussing Aboriginal people specifically. On Cape York Peninsula, the north-eastern most region of Australia through which the *Rattlesnake* expedition travelled (before reaching Papua New Guinea further north), the expedition records refer to the presence of Torres Strait Islander people (and some Papuans), as well as local Aboriginal people, among the diverse groups they encountered.

2 For more discussion on British/Aboriginal collaboration in nineteenth century exploration, see Davis 'Encountering Aboriginal Knowledge'.

3 I want to push my argument about human/environmental encounters between local Aboriginal people and British voyagers a little further, to suggest that these instances of collaboration around plants and animals and associated local knowledge might be considered as examples of what we mean in the this book by 'integrating knowledge'.

4 My thanks to Joni Adamson for making her paper on this available to me.

5 See HfE Common Threads, found on the Humanities for the Environment website: http://hfe-observatories.org/common-threads/

6 The use of the term 'bush' in the Australian vernacular has assumed an iconic status and implies a distinct understanding of what is sometimes called 'countryside'. 'The bush' refers to any sparsely inhabited region, is uniquely Australian, and is very different from the green European landscapes familiar to many European newcomers to Australia. The 'bush' in this Australian usage conjures notions of nostalgia, romanticism, and an association with a sense of heroism. The term 'The Outback' is also used, but usually in association with the more arid inland areas of Australia.

7 The violent dispossession of Aboriginal people, and the killings on the Australian 'frontier wars' in what is now the State of Queensland has been well documented by scholars (see for example Bottoms 2013).

8 Charles Card, Diary, Thursday 1 November 1849, Charles James Card Diaries, Oxley Library, Brisbane.
9 Card, Diary, Thursday 1 November 1849.
10 Card, Diary, Thursday 3 November 1849.
11 Card Diary, Thursday 1 November, 1849.
12 MacGillivray to John Gould, HMS *Rattlesnake* at Sea – 12 May 1849; Moreton Bay – 19 May 1849, in Sauer (307). London ornithologist John Gould was a sponsor of MacGillivray's collecting.
13 MacGillivray to Gould, 29[?] February 1850, from Jardine's 1850 *Contributions to Ornithology* 92–[107]. Sauer writes 'Gould authored this article entitled "A brief account of the researches in natural history of John M'Gillivray, Esq. the naturalist attached to H. M. Surveying Ship the *Rattlesnake* …". Quotes from four letters were included in the Gould article, dated May 12, 1847, February 6, 1848, May 12 & 19, 1849 and February 29, 1850' (Sauer 398).
14 Card Diary, 3 November 1849.

Works cited

'Adoption of the Paris Agreement'. United Nations Framework Convention on Climate Change, 12 December 2015, Conference of the Parties Twenty-first session, Paris, 30 November to 11 December 2015. Document FCCC/CP/2015/L.9/Rev.1 and Annex FCCC/CP/2015/L.9.

Bottoms, Timothy. *Conspiracy of Silence: Queensland's Frontier Killing-Times*. Crows Nest, NSW: Allen & Unwin, 2013. Print.

Card, Charles. Diary kept on board HMS *Rattlesnake* between 12 November 1847 and 7 March 1850. Oxley Library, Brisbane. Manuscript.

Cash, David. 'The-paris-agreement-the-first-local-global-environmental-pact'. *The Conversation*, 18 January 2016. Web.

Clarke, Philip. *Aboriginal Plant Collectors*. Kenthurst, New South Wales: Rosenberg Publishing, 2008. Print.

Davis, Michael. 'Encountering Aboriginal Knowledge: Explorer Narratives on North-east Queensland, 1770 to 1820', *Aboriginal History* 37 (2013): 29–50. Print.

Drayton, Richard. *Nature's Government: Science, Imperial Britain, and the 'Improvement' of the World*. Yale: Yale University Press, 2000. Print.

Faier, Lieba, and Lisa Rofel. 'Ethnographies of Encounter', *Ann Rev Anthrop* 43 (2014): 363–377. Print.

Goodman, Jordan. *The Rattlesnake*: A *Voyage of Discovery to the Coral* Sea. London: Faber and Faber, 2005: Print.

Ingold, Tim. *Being Alive: Essays on Movement, Knowledge and Description*. London and New York: Routledge, 2011. Print.

——*The Life of Lines*. London and New York: Routledge, 2015. Print.

Kennedy, Dane. *The Last Blank Spaces: Exploring Africa and Australia*. Cambridge, Massachusetts: Harvard University Press, 2013. Print.

Kirksey, S. Eben and Stefan Helmreich. 'The Emergence of Multispecies Ethnography', *Cultural Anthropology* 25.4 (2010): 545–576. Print.

MacGillivray, John. *Narrative of the Voyage of* H. M. S. *Rattlesnake*. Vol I. London: T. & W. Boone, 1852. Print.

McCalman, Iain. *The Reef: A Passionate History*. Melbourne: Viking/Penguin, 2013. Print.

Moore, David R. *Islanders and Aborigines at Cape York*. Canberra: Australian Institute of Aboriginal Studies, 1979. Print.

Ogden, Laura A., Billy Hall and Kimiko Tanita. 'Animals, Plants, People, and Things: A Review of Multispecies Ethnography', *Environment and Society: Advances in Research* 4 (2013): 5–24. Print.

Ogden, Laura, Nik Heynen, Ulrich Oslender, Paige West, Karim-Aly Kassam and Paul Robbins, 'Global Assemblages, Resilience, and Earth Stewardship in the Anthropocene', *Frontiers of Ecological Environment* 11.7 (2013): 341–347. Print.

Plumwood, Val. *Environmental Culture: The Ecological Crisis of Reason*, London and New York, Routledge, 2002. Print.

Rose, Deborah Bird. *Reports from a Wild Country: Ethics for Decolonisation*. Sydney: UNSW Press, 2004. Print.

Sauer, Gordon C., editor and compiler, with Ann Datta, *John Gould, the Bird Man: Correspondence, with a Chronology of His Life and Works*. Mansfield Centre, CT, USA: Maurizio Martino, In Association with the Natural History Museum, London, 2001, Vol. 4. Print.

Thoreau, Henry David. *The Writings of Henry David Thoreau, With Bibliographical Introductions and Full Indexes*, Vol IX. Boston and New York: Houghton, Mifflin and Company, 1893. Print.

13 'The Lifting of the Sky'

Life outside the Anthropocene

Tony Birch

Introduction – Goruk, the Magpie

The Indigenous owners of Country encompassing the city of Melbourne and the greater Port Phillip Bay area of the state of Victoria, Australia, comprise the Wurundjeri and the Boonerwrung. Both groups, traditionally autonomous nations in their own right, also belong to the greater Kulin nation. The Kulin's recording of its history, its understanding of the origins of Country and its reading of land and ecology is recorded through a sophisticated oral narrative, articulating an interdependent relationship between spiritual and scientific knowledge. Early European visitors and occupiers of Indigenous land too often dismissed the orally conveyed knowledge system as little more than myth or legend, with a little regard or understanding of the depth of knowledge spoken but poorly heard.

I grew up in a culturally strong urban Indigenous community, in the inner Melbourne suburb of Fitzroy. A Kulin story I first heard when I was younger, since recorded and published as the story 'The Lifting of the Sky,' went something like this:

> A long time ago there was no sky as there is now, for it lay flat upon the earth, and covered it like a blanket. It rested so hard upon it that the people were not able to move, and were in dire distress. At last Goruk, the Magpie, managed to prop up one corner of it and some of the people were freed and were able to come to his assistance. Between them they lifted it [the sky] to where it is now.
>
> (Massola 15)

'The magpie story,' as I know it, disturbed and excited me as a child. The thought that the sky had become so low to the ground and heavy that people were flattened by it frightened me. But also being a child fascinated by birds, I was drawn to the idea of the courageous magpie who managed to lift the sky and save the people. I recently became re-acquainted with the story while researching the relationship between Indigenous storytelling, science, ecology and climate change. The extent to which the story encouraged me to think about climate change was striking, when considering a sky poisoned by carbon as a result of the excessive burning of fossil fuels. The story neatly fits the definition of what Deborah Bird Rose refers to as a 'living metaphor' ('At the Billabong' 67). It is a story that transcends time. It is also a *teaching* story for contemporary society. Goruk's lifting of the sky is generally received as a 'creation story;' an explanation of the relationship between the formation of land and species, both human and nonhuman, that co-habituate Country. A Creationist interpretation locates the magpie story definitively and singularly in the past. (It also explains why some nineteenth century Christian missionaries were attracted to such stories, as, to some degree, they reflect the parables of the Old Testament.) But placing the magpie story in the past alone is a mistake. Indigenous time does not function in a linear manner. Nor does knowledge of and engagement with Country.

For Rose, and Indigenous people both, the future 'is in the ground' ('Anthropocene Noir' 8). Both past and future are also present in the sky. 'The Lifting of the Sky' is therefore also an ecocultural story, with concerns for the future clearly in mind. It provides us with a cautionary tale about climate change. It serves as a warning that if we, as a global community, continue to burn fossil fuels at current rates, or even at reduced rates that do not significantly cut the level of carbon released into the atmosphere across the planet, we will soon find ourselves in 'dire distress'. This will be as a direct consequence of the contamination of the atmosphere that produces global warming and the subsequent increase in severe and unpredictable weather events. To stop this from happening Goruk implores us to 'lift the sky' together.

Indigenous Knowledge and the Anthropocene

With climate change an issue of urgency, Indigenous Knowledge has become valued on a global scale, not only with regard to climate, but also with the ecological health and environmental balance of the planet more generally. In countries such as Canada, Traditional Ecological Knowledge (TEK) has gained legitimacy outside Indigenous communities. It has also produced cultural and ethical challenges instructive to similar relationships developing in Australia (Wilson, ed. 385–825). While TEK has historically utilised aspects of knowledge familiar to Western science, its holistic framework incorporates additional components of knowledge that science, being nominally secular, would refer to as religion. The central role of spirituality within TEK raises important intellectual and ethical issues for researchers, scientists, and environmental activists engaging with Indigenous communities. Discussion and debate of such issues can be tense, causing anger, even grief amongst Indigenous people, particularly when vested interests voice respect for Indigenous knowledge but display a tendency to 'cherry-pick' palatable portions of TEK while bypassing the vitality of the spiritual foundations underpinning knowledge. As Jon Altman discusses in *People on Country* it can be problematic that the term *traditional* itself indicates a fixed or immutable body of knowledge, when in fact local Indigenous knowledge, developed and articulated on Country is dynamic, evolving and adaptive to changing conditions. For this reason, Altman argues, it is imperative that

Indigenous communities be allowed to 'stay on country and pursue a different form of development based on working on country' (Altman 7).

The exchange of knowledge between Indigenous and non-Indigenous people can be of mutual benefit. Members of the science community have forged enriched relationships with Indigenous communities in Australia in recent years. For instance, the knowledge held in Indigenous communities with regard to climate change in the north of the country and related research conducted by both scientists and humanities scholars is producing collective ecological understanding (Tran et al.; Stacy and Tran). In future, the extent of the success of such relationships will play a key role in how we meet the challenges of climate change; and not only between Indigenous and non-Indigenous people. Climate change is an issue of human rights as much as it is of environmental or economic concern (The relationship between human rights and climate change is a key argument in both Shue and also in Nixon's *Slow Violence*). While some impoverished communities may lack access to Western knowledge in a relative sense, how they are engaged with, and valued within, a world of environmental uncertainty will become the yardstick of our collective sense of humanity. It is not idealistic to suggest that informed and genuinely equitable relationships between Indigenous and non-Indigenous communities could become a model for new forms of dialogue, practice and ethical relationships between and within other communities.

For more than a decade now a new hypothesis about 'man-made' climate change has increasingly held the attention of not only scientists, but researchers within the humanities around the 'immense, omnivorous idea' that goes by the name of the Anthropocene (Nixon 'The Anthropocene' n. pag.). A reading of the vast amount of scholarship published on the Anthropocene could lead one into believing that not only does Indigenous knowledge have no place within this discussion, but that Indigenous people across the planet have no intellectual role to play in the growing discourse. To embrace this belief would be a mistake, in large part, because there has been no serious critique of the colonial project and its devastating impact on Indigenous peoples both in Australia and globally. While recent scholarship on the Anthropocene argues for a start date coinciding with the expansion of colonialism into the Americas from the late fifteenth century, the theories surrounding 'Anthropocene discourses' needs to move beyond the superficiality of a 'collision of the Old and New Worlds' (Lewis and Maslin 174–175; Adamson 'Introduction,' this volume). The Humanities for the Environment (HfE) initiative is an international project to interrogate emerging Anthropocene discourses. This chapter seeks to interrogate, enrich and expand the ways in which these discourses are understood.[1] Within the Anthropocene narrative Indigenous nations are too often relegated to the state of *non*-existence, producing an intellectual equivalent of the *terra nullius* narrative of the late eighteenth and nineteenth centuries, a white mythology that continues to allow colonial powers to mask their histories of violence.

One definition of the Anthropocene tells us that it 'is a scientific hypothesis based on the assumption that humanity has become a global Earth system factor in sectors such as water circulation, climate, biological productivity, biodiversity and global chemical cycles' in addition to other interventions on the ecology and atmosphere of the planet (Leinfelder 9). But challenges abound in solidifying a clear understanding of what the Anthropocene actually is, and when it began, once we shift beyond initial definitions and understandings:

> It [the Anthropocene] started with agriculture 5,000 years ago, or mining 3,000 years ago. No: it starts with the genocide of 50 million indigenous people in the Americas. Or: it began with the 'Great Acceleration': the time period in the past fifty years when plastics, chemical fertilizers, concrete, aluminum and petrol flooded the market, and the environment. Or: we have no way yet to tell, we might need to wait a couple more million years.
>
> (Vansintjan n. pag.)

The Nobel Laureate Professor Paul Crutzen is credited with bringing the term 'the Anthropocene' to public attention in 2000 (Schwagerl *The Human Era*; see also Kolbert 108). Adherents of the hypothesis about a 'new epoch,' as it is badged, claim we left the period of Holocene behind some time ago, although when that time began is also the subject of debate. Regardless of the competing dates, the Anthropocene is said to mark the beginning of a new phase in the geological history of Earth, whereby human activity has permanently transformed the planet to a point that 'the world we inhabit will henceforth be the world we have made' (Purdy n. pag.). Since Crutzen's original pronouncement the hypothesis has resulted in an extensive volume of published writing by science and humanities scholars, described by the legal and environmental scholar, Jedediah Purdy, as 'a veritable academic stampede to declare that we live in a new era – the *Anthropocene* – the age of humans' (n. pag.).

A key essay in the discussion, from a humanities perspective, and one that has since been widely discussed and critiqued, is Dipesh Chakrabarty's, 'The Climate of History: Four Theses' (Chakrabarty). In 2012 a group of eminent scholars and researchers established The Anthropocene Project at the Haus der Kulturen der Welt in Berlin, Germany (Schwagerl, 'The Anthropocene'). An objective of the project is to engage researchers, academics, writers and artists from around the globe on the topic of climate change. In January 2013, the project hosted a major conference, inviting the *who's who* of the ever-evolving Anthropocene discourse. Chakrabarty delivered one of the keynote addresses to the conference, 'History on an Expanded Canvas: The Anthropocene's Invitation'.

Chakrabarty informed his audience that 'global warming is intrinsically and illogically indifferent to human injustice' ('History on an Expanded Canvas'). Just as the Anthropocene itself may be indifferent on matters of human rights and social justice, during the initial 'stampede' of scholars researching and writing about the issue a serious discussion of human rights was found wanting. Fortunately, more recently a correction has occurred, producing a critique that demands we respond in more proactive and inclusive ways to the idea of the Anthropocene. Libby Robin, for instance, the Australian environmental humanities researcher, has recently written that the Anthropocene is not only a scientific hypothesis but is also 'a metaphor for a changing society' that requires a serious consideration of the 'moral, political and ethical implications of the changes humans are making to the planet' (Robin 19).

Accepting the view that as a global community we have the capacity for genuine long-term change is a challenge of itself (albeit a necessary one), considering the extent of ecological and environmental damage suffered by the planet as a result of previous and current attempts by humans to shape it to our advantage at the expense of the Earth. It is therefore not surprising that those who accept the science underpinning the Anthropocene claim that 'the power humans wield is unlike any other force in nature' (Revkin n. pag.). Humans, however, do not act collectively as a 'force' in either a positive or a negative manner. Nor do humans impact on the planet equally, considering the inequities embedded within capitalist, colonial and even 'developing' societies. Further, there is little evidence of First World societies, those with the most destructive carbon footprint, shifting from a state of ecological vandalism to genuine environmental protection in a sustained manner. Christian Schwagerl, a founding member of The Anthropocene Project in Berlin, makes the point that the lesson of the Anthropocene will only be learned in an ecologically beneficial manner when humans finally accept that 'our civilization needs to function as an integral part of the biosphere. It's about humility towards all life, not hubris' ('How we must adjust our lifestyles to nature' n. pag.).

In consideration of human activity and enterprise underpinning the competing dates referred to as 'the golden spike' that announced the birth of the Anthropocene, often with disastrous consequences, an appeal to humility on behalf of human societies presents a daunting challenge of its own (Luciano n. pag.). Research addressing the Anthropocene's birthdate highlights the need for a wider debate on the historical and contemporary forces that have produced socio-economic disadvantage and human rights abuses as a direct outcome of environmental degradation. The global community must address the degree to which responsibly for the dire situation we face as a result of colonial and capitalist expansion should be shared by the global community. Historically, the destruction of species was an exercise of money and power, *enjoyed* by the few:

> Capitalists in a small corner of the Western world invested in steam, laying the foundation stone for the fossil economy: at no moment did the species vote for it either with feet or ballots, or march in mechanical unison, or exercise any sort of shared authority over its own destiny and that of the Earth System.
>
> (Malm and Hornborg 64)

In the influential journal *Nature*, in March 2015, Simon Lewis and Mark Maslin reviewed the existing literature on the start date of the Anthropocene. While there is evidence that humankind began to directly impact the Earth's biosystem through the early use of fire many thousands of years ago, the authors conclude that 'fires are inherently local events, so they do not provide a global GSSP'[2] and thus cannot be evidenced in earth's strata globally. Lewis and Maslin surveyed periods of human advance and expansion before focusing on four specific dates that compete for the start date of the Anthropocene. The authors deal with several periods of human history in contest for the birth of the Anthropocene. They discuss, in some detail, the 'Origins and Impacts of Farming' (beginning around 11,000 years ago); the 'Collision of the Old and New Worlds' (commencing with the arrival of Europeans in the Caribbean in 1492, and ending around 1800); the 'Industrialization' period (beginning around 1760); and finally 'The Great Acceleration' (a post-World War II phenomenon, triggered initially by the first atomic weapon detonations of 1945, followed by the widespread use of industrial chemicals) (Lewis and Maslin 174–177).

The authors conclude that only two dates ultimately complete for the start of the Anthropocene. The first is 1610, a little over a century following the initial 'collision' of 1492 in the Northern hemisphere of the Americas. The second is 'the Great Acceleration' of 1964, when the full effects nuclear fallout, resulting from earlier bomb testing, peaked. The 1610 date produced 'the unequal power relationships between different groups of people, economic growth, the impacts on globalised trade, and our current reliance on fossil fuels' (Lewis and Maslin 177). Whereas the 1964 date reflects a post-World War II rhetoric of economic and social progress underpinned by an insatiable growth in militarisation; 'Choosing the bomb spike tells a story of an elite-driven technological development that threatens planet-wide destruction' (Lewis and Maslin 177–178).

The 1610 date concludes that, as a result of the European invasion of the Americas, crop yields fell dramatically and forests were able to replenish themselves. In addition to issues of responsibility and the absence of a deeper analysis of this violence, language itself can be a problem in the reporting of research that essentially presents a case of widespread genocide all too matter-of-factly. The problematics of language are exacerbated when a highly regarded newspaper such as *The Guardian* (UK), reporting on the findings of the same research paper, concluded, in sporting parlance:

> The year 1610 marks the low-point of a dip in global carbon dioxide levels caused by a drastic reduction in farming in the Americas. This was a *knock-on effect* of the 50 million or so Indigenous deaths that resulted from the *introduction* of smallpox to the continent by European colonists.
>
> (Devlin n. pag., my italics)

The use of benign language such as this masks the brutality of colonial history for the indigenous peoples most affected, intentionally or not. The impact of colonial expansion, invasion, mass violence and its causal relationship to climate change relegates both Indigenous peoples and their lands to a status that Eyal Weizman argues ignores the direct relationship between the colonial project and climate change:

> This entangled colonial relationship to the climate demonstrates what I consider to be the most fundamental omission of the current debate around climate change. Even the most militant environmentalists still regard climate change as the 'collateral of history' – the unintended by-product of industrial development, trade and transport; whereas I see it as the intention, the very telos of the colonial project.
>
> (Prochnik n. pag.)

Reflecting Weizman's argument, the author of the influential book *Slow Violence and the Environmentalism of the Poor*, Rob Nixon, with a deliberate and refreshing flourish of emotion, states that 'the Anthropocene would have been better named the Unforgivable-crimescene' (Nixon 'The Anthropocene' n. pag.), laying bare the reality that the *anthropos*, the Greek word for 'man' or 'human,' in the Anthropocene carries a great deal of colonial baggage. This is not to suggest that contributions to the Anthropocene debate by scientists and humanities theorists are not helpful. They are. But they can also be *limiting*, absent of the necessary critique of particular historical forces that impact on any 'new epoch' thesis; not only the history of European colonial expansion, but of institutions and structures of inequality more generally, or a serious engagement with Indigenous experience of climate change historically and the ecological knowledge held in communities.

Those who favour 'The Great Acceleration' thesis are confident enough to pinpoint the start date of the Anthropocene to 16 July 1945, when the first atomic bomb was detonated in the skies above the New Mexico desert (Steffen et al. 13). One conclusion drawn from this theory on the Anthropocene is that this human made epoch is something all humans, or all *anthropos*, are equally responsible for, with an absence of an informed discussion as to who *we* are. To what extent human societies around the world should share collective responsibility for acts of environmental poisoning, such as those resulting from nuclear fallout, is dubious at best. In consideration of the relationship between a post-World War II 'Great Acceleration' and nuclear testing, it would be an absurd proposition, as Arnold and Smith suggest in *Britain, Australia and the Bomb*, that Indigenous communities living on Country in Australia should share in the responsibility for the devastation caused by the British military's testing of nuclear weapons in the South Australian desert in the 1950s; tests that poisoned people, Country and nonhuman species. Responsibility in this instance lies solely with the Australian government and the British military.

What does it mean to be human?

As introduced above, an alternate date of 1610 suggested as the birth of the Anthropocene involved the deaths of 50 million Indigenous people across the Americas. Too often in the debate, the Indigenous nations in question have been cast as little more than the *victims* of colonial expansion, reflecting again the argument forwarded by Eyal Weizman, that those who suffer the abuses of colonial violence are subsequently marginalised, forgotten, and come to exist external to a history reliant on an unproblematic narrative (Prochnik). On the subject of history, we also know that climate change is not only a future, or even contemporary event. In the past, as a result of both natural disaster and human intervention, often in the name of empire, Indigenous communities specifically, and the poor more generally, have experienced both environmental displacement and localised climate change events that have devastated communities (Nixon, 'The Anthropocene' n. pag.). As a consequence, 'what does it mean to be human?' cannot be a relative or shared question, when, historically speaking, some human societies have been provided with a greater *human* value at the expense of other societies.

The ecological and environmental challenges we face are becoming less discriminating, more visible, and increasingly impacting on all species on the planet. Donna Haraway provides the sobering observation that we need not speculate about a future event; 'system collapse is not a thriller,' she writes; 'ask any refugee of any species'. And she goes further:

> It's more than climate change; it's also extraordinary burdens on toxic chemistry, mining, depletion of lakes and rivers under and above ground, ecosystem simplification, vast genocides of people and other critters, etc., etc., in systematically linked patterns that threaten major system collapse after major system collapse after major system collapse. Recursion can be a drag.
>
> (159)

In an early critique of the Anthropocene, Eileen Crist, drawing a similar conclusion to Haraway's, was concerned that the *air* given to the Anthropocene 'detracts attention from the planet's ecological predicament as a whole, by virtue of claiming the limelight for one issue that trumps all others' ('Beyond the Climate Crisis' 38). More recently, she has also been critical of the benign language used to discuss the thesis and its relationship to history:

> How true the cliché that history is written by the victors, and how much truer for the history of the planet's conquest against which no nonhuman can direct a flood of grievances that might strike a humbling note into the human soul … the vocabulary that we are 'changing the world' – so matter-of-factly portraying itself as impartial and thereby erasing its own normative tracks even as it speaks – secures its ontological ground by silencing the displaced, the killed, and enslaved, whose homelands have been assimilated and whose lives, have indeed been changed forever; erased even.
>
> (Crist 'On the Poverty' 133)

The Country is like a body – more than wishful thinking

If we are to consider a future in which we are able to face the challenges posed by climate change in a genuinely equitable and collective sense innovative thinking (and action) will be necessary. In a recent 'thought experiment,' researchers

Stephen Turner and Timothy Neale raise the proposition that 'there is no outer settler space. There is only *someone else's country*,' Indigenous country (Turner and Neale 3). Their proposal may appear naïve, idealistic, an affront, even, to the stuck colonial mind. Turner and Neale point out that their provocation is not a frivolous act of '"wishful thinking", or an *Avatar*-like affection for indigeneity'. With climate described by the authors as 'our shared (global) endangerment' (Turner and Neale 4), they hope to shift our collective thinking, not through a radical proposal, although the idea many initially appear as so, but by one that makes ethical, intellectual and *common* sense.

Other 'experiments' when engaging Indigenous people should be approached with caution. Some non-Indigenous researchers venture 'on Country' as a means of infusing their own thinking with a concoction of 'social and sacred ecology' (Cock and Towns 23). Such an encounter with Indigenous Country might involve 'playing Indian' as some Indigenous peoples in the US have put it, or 'community work and play with the landscape followed by personal time, dinner, fire circle, and then music, drumming and dancing'. Such performances are acts of wishful thinking, at best, and could be open to ridicule. Within Indigenous communities, such initiatives raise suspicion, at the least. Such an exercise might subscribe to the over-*affection for the native* condition that concerns Turner and Neale, whereby non-Indigenous people become caught up in romantic images of 'the noble savage' with little thought given to the complexity of Indigenous culture and experience. More than this, it can produce an outcome of cultural and intellectual appropriation, whereby non-Indigenous people venture on to Country for a short-stay research trip or replenishment exercise. Indigenous people have endured these hit-and-run exercises since first contact with Europeans. Such exchanges, being all too brief and ill-conceived, benefit neither community in the long term.

In a recent essay, found in a thought-provoking collection, *Manifesto for the Anthropocene*, Anna Yeatman calls for a greater sense of *embodied awareness*, gained from being in place, physically and reflectively, as opposed to being stuck in the 'academic classroom,' literally and metaphorically. She concludes that both an idea and embodied practice 'may lead us to question what it is we think we, moderns, know, and to revalue what it is we have dismissed by way of tradition and religion' (Yeatman 125). Yeatman does not reference Indigenous knowledge or spirituality directly.[3] But her commentary does raise important issues with regard to the place of religion and spirituality in the secular world of the science and, to a large extent, the humanities research informing climate change. By inference, at least, she provokes us to consider the extent to which knowledge relationships between Indigenous and non-Indigenous communities can genuinely exist with consideration of the centrality of spirituality within the lives of most Indigenous people.

If we are to conduct what Deborah Bird Rose calls 'an ethical dialogue' between Indigenous and non-Indigenous people ('Dialogue' 129), to what degree can Indigenous spirituality be part of the conversation? And what risks are involved? Rose values risk, channelled through a more open and equitable conversation:

> One does not know the outcome. To be open is to hold one's self available to others: one takes risks and becomes vulnerable. But this is also a fertile stance: one's own ground can become destabilised. In open dialogue one holds oneself available to be surprised, to be challenged and be changed.
>
> (128)

Rose has worked closely with Indigenous communities for several decades, from the position of the *student*. In much of her writing, she repeatedly refers to her Indigenous *teachers*, with none of the paternalistic affection mentioned above. Her *place* is informed by social interaction and learning through experience, rather than by white deference, either feigned or self-flagellating. Subsequently, an open dialogue, with all the risk it involves, is, for her, rewarding. It refuses dominant models of engagement with Indigenous communities (if we can call them so), located within a 'Western critical theory and philosophical analysis' that Rose refers to as:

> A monologue masquerading as conversation, masturbation purporting to be productive interaction; it is a narcissism so profound that it claims to find a universal knowledge when in fact its violent erasures are universalising its own singular and powerful isolation.
>
> ('Dialogue' 128)

The risks of open engagement are tangible. Far more is at stake for Indigenous people within a framework of a so-called *intersubjective exchange*, a term and a performance which itself may mask monological intentions. In a society such as Australia, where Indigenous communities are subject to marginalisation and inequality at an economic and social level, any exchange with non-Indigenous people, particularly those in a position of institutional power, is likely to produce a situation of vulnerability and the potential for exploitation of Indigenous knowledge. For an *exchange* to be equitable, something of mutual value needs to be *exchanged*. Collaboration, which will become a vital framework for dealing with climate change in the future, must be an act of mutual reciprocity, in the fullest sense.

Globally, Indigenous communities have rightly demanded greater input into the debate and discussion on climate change, with regard to both the depth of ecological knowledge maintained by communities, and the specific impact that climate change and resulting extreme weather events will impact on Indigenous lives:

> Because land and sea are inextricably linked with Indigenous cultural identities, a changing climate threatens ceremony, hunting practices, sacred sites, bush tucker and bush medicine, which in turn affects law, home, health, education livelihood and purpose.
>
> (Van Neerven n. pag.)

Threats to the well-being and livelihood of Indigenous communities not only undermine the vitality and survival of communities themselves, but important cultural knowledge could also be lost, directly threatening a future of interconnected intellectual relationships both within Indigenous communities, and between Indigenous and non-Indigenous people with the common purpose of living with the planet, rather than persisting with attempts to *master* it. Whether through the extinguishment of Native Title for the benefit of mining, wanton vandalism (Birch), or forced community closures and government 'interventions' in so-called 'remote' communities (Scott and Heiss), the destruction of physical, ephemeral, spiritual and social aspects of culture denies Indigenous people connection with, and access to, ongoing customary responsibilities. Indigenous cultural survival, in the truest sense, requires a dynamic and ever-evolving engagement with Country, including whatever changes are required to adapt to shifts in ecology and environment. Adding the challenges posed by climate change to this list of interventions that currently threaten Indigenous communities means that more than *wishful thinking* is needed. Rather, co-operative and inclusive *ways of thinking* are urgently required.

Conclusion – Indigenous knowledge and the Anthropocene

The sharing of knowledge between Indigenous and non-Indigenous communities has the potential to create relationships of mutual cultural and intellectual enrichment, especially in emerging Anthropocene discourses. Such an outcome is challenging, but also possible. Remembering the initiative of Goruk, the magpie, the bird not only worked to save others from the threat of environmental destruction but also began a rescue mission that freed others, who then worked with the bird for the betterment of all. In the absence of any ethical framework able to facilitate genuine cross-cultural and intellectual exchange, the outcome of discussions about both the Anthropocene and climate change will be of negative value for Indigenous people if our perspectives and knowledges are not considered central to these discussions. It will ultimately be of negative value to outsiders also, considering the gulf that exists between genuine collaboration and white appropriation. A productive working relationship – sometimes enacted through the unstated social contract – 'you recognise my worth, I recognise yours' – has been described by the anthropologist, Eve Vincent, as an exercise in 'moral reciprocity' (Vincent 62). An important and cautionary caveat hovering over such exchanges is that Indigenous knowledge should not be sifted through for *choice*.

In Aotearoa (New Zealand), as the report 'Climate Change and the Right to Health for Maori in Aotearoa/New Zealand' indicates, the Maori have called for not only recognition of traditional knowledge, but for an active, participatory role in approaches and policies devised to deal with environmental policy generally and climate change specifically (Jones et al.). It is a role that necessitates due recognition of Maori not only as First Peoples, but as the holders of *first* knowledge. In North America, particularly in Canada, where Traditional Ecological Knowledge is taught within both Indigenous communities and the 'academic classroom,' the concept of TEK is both valued and undermined (Wilson, ed., 385–825). A call for the formal recognition of TEK by Indigenous scholars, activists and communities is accompanied by understandable hesitancy and concern, not least of all due to the structural and governmental contradictions that simultaneously debilitate Indigenous communities.

The Canadian experience offers a cautionary tale for Indigenous people globally, and particularly in Australia, where social and economic disadvantage often mirrors the North American experience. The Canadian Indigenous scholar, Waziyatawin Angela Wilson, commented a decade ago that 'before knowledge of these ways of being and interacting with the world can be shared … we must first work on recovering these traditions among our own populations' (Wilson 361-362). Another Indigenous scholar from Canada, Leanne Simpson, warns that collaboration with outsiders can result in the dilution of TEK, undermining its original objective as an anti-colonial strategy:

> The depoliticizing of Indigenous Peoples and TEK serves to make the discussion of TEK more palatable to scientists by sanitizing it of the ugliness of colonization and injustice, so scientists can potentially engage with the knowledge but not the people who own and live that knowledge.
>
> (Simpson 376)

The 'ugliness of colonisation' impacts on relationships between Indigenous and non-Indigenous people in Australia in a similar manner. Those in the environmental movement, climate change activists and researchers alike, need to examine the role of colonialism, not only in relation to the attempted destruction of Indigenous society, but the active role it has played in the warming of the planet and the accompanying increase in extreme, unpredictable and destructive weather events. Rather than avoid the tension and even conflict inherent to the establishment of new relationships, non-Indigenous people must firstly also accept the responsibility for colonial injustices. In her recent book, *Decolonizing Solidarity*, Clare Land writes 'the paternalism and tension in relationships between non-Indigenous and Indigenous activists are, at heart, generated by colonial conditions' (Land 7). Not unlike political activists, researchers within science and the humanities are subject to the same colonial conditions when dealing with Indigenous people. Time is of key importance. Time *is* of the urgency with regard to climate change. But slow time – *taking time* – is also necessary in consideration of meaningful dialogue, knowledge exchange, and long-term collective planning. Without due time being taken, without listening, without patience, teaching, learning, reciprocity is not possible. And in the absence of reciprocity, knowledge itself can die. Whether the Anthropocene is fifty or five thousand years old, what we need at present is a will to act based a respect for both difference and commonality.

Notes

1 See HfE 'Common Threads' http://hfe-observatories.org/common-threads/
2 Global Boundary Stratotype Section and Point; a measurement that indicates the presence of carbon in rock. See www.stratigraphy.org/gssp/
3 Yeatman's work is influenced by Hannah Arendt's *The Human Condition* (1958), in which Arendt meditates on what it is to be human. The human condition, Arendt argues, is designated by the term *vita active*, which is comprised of the conditions of labour, work and action.

Works cited

Altman, Jon. 'Caring for Country to Working for Country.' Eds John Altman and Sean Kerins. *People on Country: Vital Landscapes Indigenous Futures*. Sydney: The Federation Press, 2012. 1–25. Print.

Arendt, Hannah. *The Human Condition*. Chicago: The University of Chicago Press, 1958. Print.

Arnold, Lorna and Mark Smith. *Britain, Australia and the Bomb: The Nuclear Tests and Their Aftermath*. Basingstoke, Hampshire: Palgrave Macmillan, 2006. Print.

Birch, Tony. 'Nothing has changed': The Making and Unmaking of Koori Culture (1992)'. *Meanjin Anthology*. Ed. Sally Heath. Melbourne: Melbourne University Press, 2012. 227–248. Print.

Chakrabarty, Dipesh. 'The Climate of History: Four Theses', *Critical Enquiry* 35.2 Winter (2009): 197–222. Print.

——'Keynote. History on an Expanded Canvas. The Anthropocene's Invitation'. Online video clip. *YouTube*, Haus der Kulturen der Welt, Berlin, Germany. 21 January 2013. Web.

Cock, Peter and Belinda Towns. 'Wild Journeys into Social and Sacred Ecology', *PAN: Philosophy Activism Nature* 4 (2007): 21–36. Web.

Crist, Eileen. 'Beyond the Climate Crisis: A Critique of Climate Change Discourse', *Telos* 141 Winter (2007): 29–55. Print.

——'Provocation. On the Poverty of Our Nomenclature', *Environmental Humanities* 3 (2013): 129–147. Print.

Devlin, Hannah. 'Was 1610 the beginning of a new human epoch?' *The Guardian* 12 March (2015). Web.

Gibson, Katherine, Deborah Bird Rose and Ruth Fincher, eds. *Manifesto for Living in the Anthropocene*. Brooklyn, New York: Punctum Books, 2015. Print.

Haraway, Donna. 'Anthropocene, Capitalocene, Plantationocene, Chthulucene: Making Kin', *Environmental Humanities* 6 (2015): 159–165. Web.

Jones, Rhys, Hayley Bennett, Gary Keating and Alison Blaiklock. 'Climate Change and the Right to Health for Māori in Aotearoa/New Zealand', *Health and Human Rights Journal* 16.1 July (2014). Web.

Kolbert, Elizabeth. *The Sixth Extinction: An Unnatural History*. London: Bloomsbury, 2014. Print.

Land, Clare. *Decolonizing Solidarity: Dilemmas and Directions for Supporters of Indigenous Struggles*. London: Zed Books, 2015. Print.

Leinfelder, R. 'Assuming Responsibility for the Anthropocene: Challenges and Opportunities in Education'. Ed. Helmuth Trischler, 'Anthropocene: Exploring the Future of the Age of Humans', *RCC Perspectives* 3 (2013): 9–28. Web.

Lewis, Simon L. and Mark A. Maslin. 'Defining the Anthropocene', *Nature* 519, 12 March (2014): 171–180. Print.

Luciano, Dana. 'The Inhuman Anthropocene', *Avidly*, 22 March (2015). Web.

Malm, Andreas, Alf Hornborg. 'The geology of mankind? A critique of the Anthropocene narrative', *The Anthropocene Review* 1.1 (2014): 62–69. Web.

Massola, Aldo. *Bunjil's Cave: Myths, Legends and Superstitions of the Aborigines of South-East Australia*. Melbourne: Lansdowne Press, 1968. Print.

Nixon, Rob. 'The Anthropocene: The Promise and Pitfalls of an Epochal Idea', *Edge Effects*, 6 November (2014). Web.

——*Slow Violence and the Environmentalism of the Poor*. Cambridge, Massachusetts: Harvard University Press, 2011. Print.

Prochnik, George. 'George Prochnik Interviews Eyal Weizman. The Desert Threshold', *Los Angeles Review of Books*, 18 October (2015). Web.

Purdy, Jedediah. 'Anthropocene fever. The Anthropocene idea has been embraced by Earth scientists and English professors alike. But how useful is it?' *Aeon*, 31 March (2015). Web.

Revkin, Andrew C. 'Did Earth's "Anthropocene" Age of Man Begin With the Globalization of Disease in 1610?' *International New York Times*, 11 March (2015). Web.

Robin, Libby. 'A Future Beyond Numbers', *Welcome to the Anthropocene. The Earth in Our Hands*. Eds. Nina Möllers, Christian Schwägerl and Helmuth Trischler. Munich: Deutsches Museum, 2015. 19–24. Print.

Rose, Deborah Bird. 'Anthropocene Noir', *People and the Planet 2013 Conference: Transforming the Future*, RMIT University, Melbourne, Australia, 2013. Web.

——'At The Billabong', *PAN: Philosophy Activism Nature* 5 (2008). Web.

——'Dialogue', *Manifesto for Living in the Anthropocene*. Eds Katherine Gibson, Deborah Bird Rose and Ruth Fincher. Brooklyn, New York: Punctum Books, 2015. 127–131. Print.

Schwagerl, Christian. *The Anthropocene: The Human Era and How it Shapes Our Planet*, Santa Fe, New Mexico: Synergetic Press, 2014. Print.

——'How we must adjust our lifestyles to nature: Welcome to the "Anthropocene", the human epoch'. *Independent*, 25 February (2015). Web.Scott, Rosie and Anita Heiss, eds. *The Intervention: An Anthology*. Sydney: Concerned Australians, 2015. Print.

Shue, Henry. *Climate Justice, Vulnerability and Protection*. Oxford: Oxford University Press, 2014. Print.

Simpson, Leanne R. 'Anticolonial Strategies for the Recovery and Maintenance of Indigenous Knowledge', Special Issue: 'The Recovery of Indigenous Knowledge.' Ed. Waziyatawin Angela Wilson. *The American Indian Quarterly* 28.3&4. Summer/Fall (2004): 373–385. Print.

Stacey, Claire and Tran Tran. *Climate Change Adaptation on Karajarri Country, Community Report*. Acton, ACT: Australian Institute of Aboriginal and Torres Strait IslanderStudies, 2013. Web.

Steffen, Will, Wendy Broadgate, Lisa Deutsch, Owen Gaffney and Cornelia Ludwig. 'The Trajectory of the Anthropocene: The Great Acceleration', *The Anthropocene Review* 16 January (2015): 1–17. Web.

Tran, Tran, Lisa Strelein, Jessica Weir, Claire Stacey and Anna Dwyer. *Native Title and Climate Change. Changes to Country and Culture, Changes to Climate: Strengthening Institutions for Indigenous Resilience and Adaptation*. National Climate Change Adaptation Research Facility, Gold Coast, Queensland, 2013. Web.

Turner, Stephen and Timothy Neale. 'First Law and the Force of water: Law, Water, Entitlement', *Settler Colonial Studies* 5.4 (2015): 1–11. Web.

Van Neerven, Ellen. 'The Country is Like a Body', *Right Now* 26 October (2015). Web.

Vansintjan, Aaron. 'The Anthropocene Debate: Why is Such a Useful Concept Starting to Fall Apart?' *Resilience* 26 June (2015). Web.

Vincent, Eve. 'Hosts and Guests: Interpreting Rockhole Recovery Trips'. *Australian Humanities Review* 53 November (2012): 61–78. Web.

Wilson, Angela Cavender. 'Introduction: Indigenous Knowledge Recovery Is Indigenous Empowerment', Special Issue: 'The Recovery of Indigenous Knowledge.' Ed. Waziyatawin Angela Wilson. *The American Indian Quarterly* 28.3&4. Summer/Fall (2004): 359–372. Print.

Wilson, Waziyatawin Angela, ed. Special Issue: 'The Recovery of Indigenous Knowledge.' *The American Indian Quarterly* 28.3&4. Summer/Fall (2004). Print.

Yeatman, Anna. 'The Human Condition in the Anthropocene'. *Manifesto for Living in the Anthropocene*. Eds Katherine Gibson, Deborah Bird Rose and Ruth Fincher. Brooklyn, New York: Punctum Books, 2015. 123–126. Print.

14 Literature, ethics and bushfire in Australia

Kate Rigby

When the Australian Working Group for the Ecological Humanities first convened at the start of the new millennium at the Australian National University's Centre for Resource and Environmental Studies, it was as a small group of scholars from different disciplines, each of whom had developed their own angle on the contribution of the humanities and social sciences to diagnosing and redressing the socio-cultural dimensions of the environmental problems that weighed upon us. The shared vision that emerged from these early discussions was encapsulated in a manifesto that framed the Ecological Humanities as an interdisciplinary and intercultural endeavour, dedicated to 'rethreading the fabric of knowledge' by 'building bridges' between 'the sciences and the humanities, and between western and other ways of knowing,' with a view to 'developing moral action in relation to the "natural" world'. This undertaking, we affirmed, was motivated by curiosity, uncertainty, concern, and a desire to collaborate with 'scholars and other experts from a diversity of cultures and traditions,' along with a commitment to 'cultural, biological, and academic diversity' (Rose and Robin 'Manifesto'). While it was not made explicit here, the scare quotes around 'natural' signalled our keen awareness of both the cultural construction of concepts of 'nature' and of the interrelationship of 'social and ecological justice' (Rose and Robin 'Ecological Humanities in Action').

Over the previous decades, ecologically oriented approaches and methodologies had been developed in each of the disciplines represented in this working group, with my own, ecocritical literary studies, being the new kid on the block by comparison with ecophilosophy, environmental history and the kind of multi-species ethnography (which was yet to be named as such) that Debbie Rose had been developing through her long-standing work with some of Australia's First Nations peoples and cultures. These green threads nonetheless remained minor strands within our respective disciplines, and most of us felt more-or-less isolated within our institutional location: either as the only ecologically engaged researcher in the department, or, in the case of Rose and Robin, as the only humanities scholars in a centre dominated by natural scientists. Tellingly, perhaps, the first publishing outlet that we found for our collective enterprise was modestly called a 'corner'. Since then, this minority status has begun change, although not as much as we might have wished. As Castree et al. observed in 2014 in their feisty perspective article in *Nature Climate Change*, the capacity of the Environmental Social Sciences and Humanities (ESSH) to help analyse and resolve those environmental problems, including the notoriously 'wicked' one of fossil-fuelled climate change, which arise from socio-cultural and political-economic no less than from techno-scientific factors, continues to be undervalued in the arena of global environmental change science and policy development, which consequently operates with a 'stunted' conception of their 'human dimensions' (763).

The very publication of this perspective on 'Changing the Intellectual Climate' is nonetheless indicative of the recent efflorescence of the Environmental Humanities (as we too in Australia had somewhat reluctantly come to call our enterprise and its flagship journal), the growing recognition of which, both within and beyond the academy, was both evidenced and advanced by the Mellon Foundation's generous funding of the Humanities for the Environment Observatories (2013–2015). In the meantime, we had also honed the ways in which we articulate our enterprise. This can be seen, be for example, in the co-authored article in *Trends in Ecology and Evolution* from 2007, which arose from a multi-disciplinary symposium at the ANU's Fenner School the previous year, addressed to the necessity of bridging the divide between the 'natural' and 'human' sciences in order to close the 'growing "sustainability gap" between what we know needs to be done and what is actually being done' (Fischer et al. 621). Observing that '[h]uman action in the world emerges from a complex dialectic among the living world itself, the social contexts of human life and action, and the conceptualisations through which human life is made meaningful,' the authors of this landmark article argue that '[f]undamentally enhanced collaboration among natural and social scientists and scholars of human contexts, symbols and meanings would signal the beginning of a new paradigm for addressing the sustainability gap' (623). One of the things that makes this article so significant, in my view, is that its co-authorship, which included Rose and Robin alongside ecologists, conservation biologists and physical geographers, itself instantiates the kind of cross-disciplinary conversations that it calls for. Strategically placed in a science journal, it is also indicative of a growing readiness among some humanities scholars to get outside their disciplinary comfort zone in order to have their voices heard in the world of the physical sciences. This endeavour is taken further by Castree, in conjunction with his sizeable cohort of contributors, in the aforementioned Perspective piece, which concludes with an exhortation to EH and ESS scholars 'who feel they are not part of the "GEC conversation" beyond their home discipline' to 'break in to the relevant meetings, conferences and journals' (767). Appearing in a very high profile international scientific journal, this article also provides a more detailed definition of those underlying 'human dimensions' of global environmental change, which are not adequately grasped within current research initiatives that exclude the humanities and qualitative social sciences. These include:

> questions of value, responsibility, rights, entitlements, needs, duty, faith, care, government, cruelty, charity and justice in a world marked by (1) significant differences in people's customs and aspirations, (2) manifest

> inequalities in people's living conditions and material prospects, and (3) complex material and moral interdependencies among people and non-humans stretched across space and unfolding through time. [...] The environmental humanities illuminate peoples' complex and divergent understandings of life – human and non-human – on Earth. They also pay close attention to human faculties beyond cognition and reason, dealing with such things as love, trust, fear, care, commitment, devotion and loyalty. (765)

Working across the methodological differences that rightly characterise distinct humanities disciplines, moreover, EH researchers have also started to pay closer attention to the ways in which these might be productively integrated with respect to particular locations and matters of concern (Hartman).

My own research is located on the 'common ground' between environmental history and ecocritical literary and cultural studies mapped by Bergthaller et al., in their 2014 article in *Environmental Humanities*, while also drawing on ecophilosophy, ecotheology and eco-hermeneutical biblical studies. Additionally, and crucially, it is impelled by an awareness that the global environmental, or, as I prefer, socio-ecological, crisis that our initial working group had optimistically thought the Ecological Humanities might help to avert, is already upon us. It is frequently remarked that that those (both human and otherwise) who have hitherto been worst affected by this unfolding crisis tend to be among those who have done least to cause it. While this is doubtless true, it was a calamity that bore down upon a centre of human power (nationally) and privilege (globally), namely Australia's federal capital, that first suggested to me the urgent necessity of learning how to live, justly, compassionately, and sustainably, with heightened levels of risk and uncertainty in an era of potentially calamitous environmental and climatic change. Witnessing a tornado of flame descending upon Canberra on 18 January 2003, I was struck by the ill-preparedness of even so well-heeled and well-educated a town as this for the increasingly calamitous conditions that our anthropogenically altered climate was engendering. This ill-preparedness, I realised, has affective, ethical and conceptual dimensions, pertaining to deep-lying cultural assumptions about, and feelings towards, the more-than-human world. It was with a view to unearthing such problematic assumptions, and advancing alternative ways of thinking, that I turned my attention to the eco-cultural history and literary representation of those calamities that have come to be known in modernity as 'natural disasters'. In the book that arose from this research, *Dancing with Disaster*, I focussed especially on the capacity of narrative fiction to provide an ethical space of reflection disclosing the dynamic interrelationship of human and nonhuman agencies and processes in the unfolding of such calamities. Here, however, I conclude with (an all too brief) discussion of a remarkable poetic text by Melbourne author Jordie Albiston, which draws upon the biblical language of prophetic lament in response to Victoria's 'Black Saturday,' the most recent and thus far most intense of Australia's mega-blazes.

A large part of the problem, in my analysis, lies with the very framing of such calamities as 'natural disasters'. Taking shape from the mid-eighteenth century, but entering the English lexicon only towards the end of the following century, this term arises from what Michel Serres (31–32) has termed the Modern Constitution that became embedded in the institutionalisation of knowledge production during the nineteenth century, engendering that very severance of the 'natural' from the 'human' sciences that the Environmental Humanities has set out to mend. This dualistic onto-epistemology took shape and struck root at precisely the time when science and technology were enrolling ever more nonhuman entities into the complex networks that are constitutive of industrial societies. Along with the things that most of us value, such as refrigerators and motor cars, this process has unintentionally generated sundry seriously undesirable natural–cultural 'hybrids,' as Latour (108) calls them, such as ozone depletion and global warming. Meanwhile, coming to terms with these hybrid phenomena, which can only be understood and redressed through a transdisciplinary lens, has been rendered all the more difficult because of the great divide between the sciences and the humanities, along with the marginalisation of alternative knowledge systems (notably, those of colonised peoples). Similarly, the crucial task of recognising and addressing the multiple human and nonhuman agencies and processes that go into the making of disasters such as the firestorm that consumed part of Australia's federal capital in 2003 has been severely hindered by the modern notion of 'natural disaster'. Ironically, the very period when long-term human impacts upon the planet have escalated to such an extent that our era has been dubbed the Anthropocene is also the era in which the entanglement of morality and materiality, social relations and natural phenomena, has become veiled.

As the editors of a volume on the anthropology of disaster from 1999 observe,

> disasters spring from the nexus where environment, society, and technology come together – the point where place, people, and human construction of both the material and nonmaterial meet. It is from the interplay of these three planes that disasters emanate, and in their unfolding, they reimplicate every vector of their causal interface.
>
> (Hofmann and Oliver-Smith 1)

Generally framed as dramatic events, disasters are better understood as material-discursive processes, in which the varying needs and interests of the individuals, groups, and organisations affected are 'articulated and negotiated over the often extended duration of the entire phenomenon' (ibid. 12). Occurring in the context of longer-term processes of socio-ecological development – or mal-development – even those calamities that are triggered by non-anthropogenic phenomena frequently participate in the inequitable distribution of environmental harms that Rob Nixon has termed 'slow violence'. No such disaster, then, is ever purely 'natural,' but always to a greater or lesser extent socio-ecological: as such, they are better described as 'eco-catastrophes' (Carrigan). In the face of the increasing frequency and intensity of weather-borne extremes, moreover, the continued designation of such calamities as 'natural disasters' is particularly maladaptive. For it conjures a cultural narrative that effaces the anthropogenic dimension of the calamitous conditions that now confront us, fostering a fearful and hostile attitude towards a reified 'nature' at the very time when we most need to better appreciate the connectivities, both material and moral, linking human wellbeing with that of other living beings and with those volatile bio-physical systems that both enable and, at times, endanger our collective flourishing.

In Australia, it was in the wake of the catastrophic Victorian firestorm of 2009 that debate really began to rage about the interpretive framing of those weather-borne eco-catastrophes that are now proliferating around our perilously warming planet. On 7 February, at the height of an almost decade-long drought and in the midst of an unprecedented heatwave, several of the 400 fires burning across Victoria converged, obliterating 6 small towns and badly damaging many more, taking 173 human lives, destroying 2,133 homes and displacing over 7,500 survivors (Teague et al.). Over 450,000 hectares of land were burnt out, killing or injuring countless domestic and free-living animals, and blitzing 40 per cent of the remaining habitat of Victoria's faunal emblem, the endangered Leadbeater's Possum (Lindenmeyer et al.). When ecophilosopher Freya Mathews sought to amplify the voice of eminent Australian climate scientists by writing an opinion piece in *The Age* that framed 'Black Saturday,' as 'the face of climate change in our part of the world,' she was subjected to a torrent of abuse. Mathews' eagerness to interpret this eco-catastrophe through the lens of climate change might have obscured another of its key lessons: namely, the continuing failure of most Australians to reckon with the 'holocaust fires,' as Tom Griffiths terms them ('Unnatural Disaster?'), which are endemic to Victoria's majestic Mountain Ash forests, and indeed essential for their renewal (although preferably only once every 300 years or so). Meanwhile, Australia's sole national newspaper, the Murdoch-owned *Australian*, led a veritable campaign to undermine any association between global warming and the extended drought and unprecedented heatwave that culminated in this catastrophe, targeting instead those environmentalists who had allegedly prevented the widespread prescribed burning, which, it was claimed, would have stopped the fires from becoming so ferocious. Miranda Devine took this fight against 'eco-terrorism' to *The Sydney Morning Herald*, where she insisted that 'it is not arsonists who should be hanging from lamp-posts but greenies'.

It may well be that this particular campaign was being prosecuted, at least in part, by those with a vested interest in obfuscating climate science and derailing mitigation efforts. But this narrative was only able to gain traction with the public, in my view, because the traumatic experience of the fires had stirred up that underlying, culturally-conditioned contempt, fear, and even hatred towards the natural environment which Simon Estok has termed 'ecophobia': an ecophobia that could readily be directed towards those traitorous 'greenies' whose 'love of nature' was construed as a betrayal of their own kind. In the more sober analysis of the *2009 Victorian Bushfires Royal Commission Report*, more extensive preventive burning is indeed included among the recommendations; but so too are improvements to the electricity distribution system, since five of the nineteen mega-fires that swept the State that day were attributed to the failure of electricity assets. Improved building codes, better warning systems, and more extensive public education about living in the vicinity of endemically fire-prone bushland were also strongly endorsed. In a further echo of the condemnatory conclusions of Leonard Stretton's High Commission Report on the Black Friday Victorian firestorm of January 1939, fifteen of the remaining fires were determined to have been the direct or indirect result of human actions, while four were thought to have been lit by arsonists. Climate change, however, was notable by its absence from this this report.

In *Dancing with Disaster*, I turned to the work of environmental historians in order to put Black Saturday into a wider perspective. One of those was Stephen Pyne, from whom I learnt that, on this peculiarly fiery planet – the only one in our solar system in which the evolution of organic life-forms has engendered just the right conditions for free-ranging flame – the Australian continent boasts a biotic community that has been more decisively refashioned by fire-wielding humans than any other in the world. The use of fire is humanity's primal technology, and it has played a critical role in the development of all human cultures and the colonisation of new climes. And when humans began to dance with fire, other beings got caught up in the whirl. Wherever humans ventured, the life-cycles of plants and animals became attuned to the rhythms of anthropogenic fire, while those who could most readily adapt to this new biotic ballet fared better than those who could not. In many parts of the world, but nowhere more so than on the continent that the British dubbed Australia, multi-species collectives gradually came into being through what I have called a process of 'pyro-symbiogenesis,' which required the element of anthropogenic fire for maximal flourishing. As a protean element that acquires the properties of weather, wind, terrain and fuel, however, fire is not a tool in the conventional sense, and it has always been an unruly companion, never totally at the command of even the most skilled human practitioners. The new kinds of agency that humans acquired by entering into alliance with what Catriona Sandilands has dubbed this 'queer' companion were therefore far from risk-free, and brought with them a new burden of responsibility. Fire could be engaged carefully and skilfully in ways that favoured human sustenance, while benefitting also those other-than-human lives with which ours was entangled; or it could be wielded heedlessly 'like the spray cans of environmental vandals' (Pyne 'Consumed by Fire' 82). The use of fire might be an anthropological constant; but human fire cultures and histories vary significantly.

How Australia became so very fiery is an exemplary tale of pyro-symbiogenesis. In environmental historian Tom Griffiths' account (*Forests of Ash*), the first long, slow move in this dance of distributed agency was made by the lithosphere: when the super-continent of Gondwana broke up around 45 million years ago, the segment that became Australia drifted north towards the Tropics, getting hotter and dryer along the way. Ice Ages came and went; earthquakes and volcanoes altered the face of the land; species evolved and disappeared, and when a land-bridge was formed to the smaller islands to the north, new ones arrived. Among them, eventually, were human beings. By the time they got here, almost certainly by boat, and probably around 50,000 years ago, many Australian plants had already learnt to dance with fire. Only a small percentage of the ancient Gondwanan species survived in the tropical rainforests of the north and in remnant temperate rainforest regions of the southeast. Throughout most of the continent, plants had adapted to increasing aridity by developing small, hard and sometimes spiny leaves ('scleromorphy'), while the trees that predominated throughout Australia's sclerophyll forests and grassy woodlands were eucalypts: 'fire weeds,' Pyne terms them (*World Fire* 31), with no disrespect intended, but because most have evolved an extraordinary capacity to withstand fire, and all require it in order to germinate. 'During Greater Australia's lonely latitudinal drift, the continent became embraced by fire just as its abandoned partner, Antarctica, loitering at the pole, became overwhelmed by ice' (*Forests of Ash* 5).

The nature and extent of the environmental impact of the early human colonisation of this continent remains a matter of heated debate, but it currently appears unlikely that Aboriginal peoples were wholly responsible for the disappearance of any or all of Australia's megafauna, for if their hunting practices did play a part, so too did climatic changes (Paterson 4–5). What is without doubt, however, is that the first Australians ultimately got the hang of living on this largely arid, fire-prone continent. Having survived the last glacial maximum, they honed their mobile and flexible way of life in response to a labile climate, characterised by frequent extremes, low predictability and non-annual cycles. Most of the continent's megadiverse plants and animals, moreover, learnt to dance along with the sophisticated fire regimes that they established. The outcome of this process of pyro-symbiogenesis was not only a peculiarly pyrophitic biota, but a human society that had become extraordinarily adept at what Rhys Jones famously termed 'fire-stick farming'.

By contrast, the culture of the overwhelming majority of the new wave of people who began to colonise Aboriginal country from the late eighteenth century was shaped by environmental conditions that were uncommonly inhospitable to free-ranging fire. It was only in the company of fire that modern humans, like Neanderthals before them, had been able to colonise the cooler climes of northern Europe. During the Enlightenment, however, urban agricultural experts advocated the cessation of the one use of free-burning fire that had remained common on the land. Faced with the necessity of increasing food production for a growing population, and committed to putting agriculture on a rational basis, these would-be 'improvers' condemned traditional field rotation and fallow-firing practices as archaic and wasteful. Over the course of the eighteenth century, free-burning fire became associated with itinerancy, backwardness and social unrest. Fire belonged in the hearth, not on the ground. Notwithstanding the year-round sogginess of northern European lands, this intensification of crop production would probably have impoverished the soil had the suppression of crop rotation and fallow-firing not been compensated by the introduction of new fertilising agents appropriated from other places and times: guano from the Peruvian islands, and, when supplies of that began to dwindle, phosphates extracted from pulverised rock. The replacement of ash and dung by superphosphates was part of that much bigger shift whereby the controlled burning of flammable materials harvested on the Earth's surface was, as Pyne puts it, 'sublimated' into the controlled combustion of ancient biomass extracted from beneath the ground (and, later, the sea) ('Consumed by Fire', 91).

Notwithstanding their pyrophobic proclivities, however, many of these new Australians soon took to setting fires in the open: itinerant 'swagmen,' following paths that had been made over countless generations by Aboriginal feet, lit 'billy' and camp-fires wherever they stopped along the way; 'squatters' (a term originally coined to refer to those who illegally appropriated grazing land beyond the official line of British settlement, but subsequently used for graziers in general) learnt to improve their pasture by seasonal burning; 'selectors' on the smallholdings made available by government policies to encourage closer settlement in the 1860s used fire, in conjunction with the axe, to open up the 'scrub' for farming and then to burn off the stubble after the harvest; and miners used fire to gain access to prospecting sites. In those parts of the country where the temperate climate made closer settlement a realistic venture, Aboriginal fires were gradually but surely extinguished during the course of the nineteenth century, as the firestick was wrested into colonial hands. Compared with Aboriginal fire, however, colonial fire was lawless, and generally ill-fitted to existing ecosystems and climatic conditions.

While fire is endemic to the Australian bush, including, in the case of Victoria's Mountain Ash, monumental wildfires, 'Black Saturday' 2009 is the most recent of the mega-blazes, continental 'fires of regime change,' as Pyne terms them (*The Still-Burning Bush* 33), which followed the unsettling engendered by European colonisation, and which now appear to be becoming more frequent under the global regime change of planetary warming. These mega-blazes have predominantly affected the south-east, where topography and winds, together with long dry periods and fire-hungry forests, conjoin to engender those terrifyingly fast-moving, sky-darkening tornadoes of flame known as firestorms. The first, 'Black Thursday,' which also affected large swathes of Victoria, arrived on 6 February 1851. It coincided with the start of the Gold Rush and was linked, at least in part, to the disruptive impacts of grazing. It was followed by 'Red Tuesday,' 1 February 1898, which blazed in the Otway and Dandenong Ranges in the wake of decades of ring-barking and clearing. Smoke from these fires drifted into the Victorian Parliament, discomforting the assembled statesmen as they argued over the details of Australian federation. 'It seems appropriate,' as Richard Evans observes (2009: 106), 'that the Australian nation was shaped in this setting: the founding fathers in their three-piece woollen suits gasping for breath'. Next up, again in Victoria, was 'Black Sunday,' 4 February 1926, which originated in forests reserved for timber harvesting, and swept through Gippsland, the Dandenong Ranges, Yarra Valley and Kinglake area. According to Pyne, however, all of these blazes 'paled before the Black Friday fires of 1939 that seemed to sweep all the wreckage and violence of settlement into one colossal maelstrom' (*World Fire*, 38). In Victoria on 13 February of that year, a massive firestorm burnt out around 1.4 million hectares, destroying nearly five million cubic metres of timber in the central highlands alone, devastating several towns, killing countless animals, both wild and domestic, along with seventy-one people, and leaving many more injured, traumatised, and homeless (Griffiths *Forests of Ash*, 134). Ash from this inferno fell across the sea in Tasmania and even New Zealand.

Firestorms such as these were utterly outside the experience of settler Australians, who also had to contend with the smaller, more localised, but nonetheless potentially devastating bushfires that became a virtually annual occurrence, especially during dry periods. Fire acquired emblematic status in colonial Australia, alongside drought and flood, as malign forces against which hardy settlers were called upon manfully to do battle in their conquest of the wayward colonial earth. Their struggle with the elements, above all with the ever-present threat of summer bushfires, formed the troubled terrain on which colonial Australians conceived the egalitarian ideal of 'mateship'. The utopian impulse underlying this cultural formation can be discerned in texts such as 'The Fire at Ross's Farm' (1891) by one of the great Australian balladists of the late nineteenth and early twentieth century, Henry Lawson (1867–1922). What initially looks like an Antipodean version of the tragic tale of Romeo and Juliet, to which the singer explicitly alludes, turns into a narrative of male bonding across class lines, as Sandy the Scottish squatter finally rides up with his men to join the selector, Ross, with whom he has long been at odds, and his own son, Robert, who is enamoured of the humble selector's daughter, in protecting the smallholding of his erstwhile enemy:

> And when before the gallant band
> The beaten flames gave way,
> Two grimy hands in friendship joined-
> And it was Christmas Day.
>
> (40)

The blessings of Christmas invoked in the closing lines of this ballad celebrate a human community created in the shadow of disaster. But the bounds of this community are very narrowly defined. Class differences, it is true, have melted away in the heat of the blaze. But this egalitarian community is exclusively masculine and white: 'pretty Jenny Ross' features only as the prize for which the squatter's son is fighting, while the total absence of Aboriginal figures is not even meliorated by the recollection that it was Indigenous landholders who had inducted settler Australians into the use of those green branches to quell fires. Moreover, while Lawson does spare a thought for some of the nonhuman victims of the fire – the bees that 'fell stifled in the smoke/Or perished in their hives,' and the stock that joined the kangaroos in 'flying for their lives' – the world of white, male mateship that he invokes here is only brought into being by means of the identification of a common enemy in the pyrophitic proclivities of the colonial earth. Lawson's anthropo-, andro- and ethnocentric community of equals is created by casting wildfire as a primordial foe:

> Like sounds of distant musketry
> It crackled through the brakes,
> And o'er the flat of silver grass
> It hissed like angry snakes.
> (40)

This mythos of the 'Aussie battler,' struggling to make a life for himself in a land of promise, but beset by droughts, fires, and floods (along with poor soils, foot-rot, and sundry plant and animal pests), retains a strong hold over the Australian social imaginary. During Australia's 'Angry Summer' (Steffen) of 2013, for example, it was invoked by Australia's then Prime Minister, Julia Gillard. Expressing her condolences to the families of those who had died in the unfolding Queensland flood disaster while visiting others who were afflicted by bushfires in Victoria, Gillard declared on ABC's Lateline programme (Barlow) that the whole country was being 'challenged by nature,' but that 'we are a strong and smart nation and we'll get through this, as we always do, by pulling together'. Gillard's reassurance suggests that it is once more us Aussie battlers holding out against the unruly elements; and it is our (exclusively human, although now more socially inclusive) 'mateship' and superior smarts that will get us through it. Potentially fuelling ecophobia, this response also militates against the recognition of these extremes as something other than what settler Australians have experienced in the past: namely, as at least partially anthropogenic, and hence as 'sentinel events,' in which we should be discerning the wrongs of our current modus vivendi, a foreboding of worse to come, and an urgent call to change our ways.

While omitting any explicit mention of climate change, Jordie Albiston's extraordinary poem 'Lamentations' can be read as implicating just such a call. Seeking a language of response to the horrendous damage and loss of life sustained on Black Saturday in Victoria's Kinglake area, where 119 of the 173 human fatalities occurred (and where Albiston herself had formerly resided for much of her life), she sought poetic inspiration from the biblical 'Book of Lamentations,' attributed to the prophet Jeremiah. While this authorial ascription is now thought by biblical scholars to be incorrect, the placement of the biblical 'Lamentations' directly after the Book of Jeremiah frames it in such a way as to lend a prophetic dimension to this series of songs, which were composed (probably by more than one author) in response to the Babylonian sacking of Jerusalem in 586 BCE. Jeremiah was profoundly critical of his own society, targeting in particular the treacherous trading practices that had allowed the ruling elite to grow 'great and rich [...] fat and sleek' (Jer. 5: 23–24) at others' expense, while failing to 'protect the cause of the orphan' and 'defend the rights of the needy' (Jer. 5: 27–28). In addition, Jeremiah laments their indifference to the drying up of the land (troped as 'mourning') and the disappearance of 'birds and animals' (Jer. 12.4), alluding to a regional environmental crisis that he also associates with human wrong-doing: 'They made it a desolation,' cries the Lord through the voice of the prophet, 'desolate it mourns to me. /The whole land is made desolate,/but no one lays it to heart' (Jer. 12:11). Read in conjunction with Jeremiah's searing socio-ecological critique, then, the succeeding songs of sorrow for the destruction of Jerusalem become linked with a grief-stricken awareness of the socio-political ills that had beset the city prior to the Babylonian invasion, which thereby acquires the force of a sentinel event.

Bewailing the destruction of Kinglake and the devastation of the surrounding bushland, Albiston follows the Hebrew original of the first four poems of the 'Lamentations' in composing her verses as alphabetical acrostics, stopping each time at the letter V: V, that is, for 'Victoria,' at the conclusion of the first three poems in the sequence (I–III), and for 'vanquished' in the last (IV). The incompletion of this alphabetic series recalls the alphabet of the Hebrew original, which has only twenty-two letters; but it can also be read as an acknowledgement of the inability of words to convey the full horror of this event. Crucially, Albiston also replicates the reversal of the letters P and Q in poems II, III and IV, to indicate that *all is not right*. Echoing other formal features of the 1611 Authorised Version of the English translation, such as the seemingly arbitrary use of italics (actually marking words supplied by the translators for grammatical purposes), the sense of local detail, and general desolation as expressed through specific imagery, the music of rhythm and repetition, and a formal register of communal grief,[1] Albiston nonetheless departs from the biblical 'Lamentations' in certain key ways: firstly, by carefully excluding any suggestion of divine vengeance; and secondly, by emphatically including diverse nonhuman others among the innocent victims of the firestorm:

> *I.*
>
> Ah, look how the township sits solitary *that* was so full of people: look how she sits like a weeping widow, the town that just yesterday! sat queen of Murrindindi, *and* of the Great Dividing Range, that sat jewel in the crown of all Melbourne.
> Black is the only one here: black is the only one left: whichever way we turn it is black who meets our eye, black who shakes our hand, black who murmurs nothing in our ear.
> Can you believe it?
> Do you believe what you see?
> Everything is missing now, there is no movement in the bush: everything is gone and there is no bush.
> Flora, fauna, family.
> Gone.
> How has it all come to this?
>
> (35)

While there is no definitive answer to this question, in the third poem the speaker alludes to a number of actors and factors, both human and otherwise that contributed to the catastrophe:

> *III*
>
> [...]
> Understand the story of firestorm and flame, of north wind and southerly change.
> Understand the story of drought and of fuel.
> Understand, and understand: and understand again.
>
> (40)

In an anthropogenically warming world, to 'understand the story of drought and fuel' is to recognise not only the necessity for preventive burning to reduce the 'fuel' that fans the flames on days such as this; it is also to acknowledge that the day-trippers, who had been introduced in the retrospective second poem set just prior to the firestorm, are among those whose carbon dioxide emissions, the 'fuel' that powers their cars and so much else in their lives, have inadvertently contributed to the unprecedented heat wave that created the conditions in which this firestorm took hold:

> *I.*
>
> Ah, today you are bringing your baskets to Kinglake.
> Bread and meat are in your baskets: you are bringing them to Jehoshaphat Valley, *and* to Masons Falls, and to Kinglake National Park.
> Children are in your cars: you are bringing them with your biscuits to picnic in Kinglake.
> Driving up the mountain, you stop to snap shots of flowers, and of the forest, and of Strathewen far below: and of Melbourne, farther below, *that* is covered with cloud in its heat.
>
> (36)

Valuing 'nature' aesthetically, happily snapping photos while failing to register either the disruptive impact of their fossil-fuelled lifestyle or the warning signs of immanent fire-danger, such privileged day-trippers are ironically hastening the devastation of those very places that they like to visit recreationally on sunny Sundays. That devastation, Albiston stresses, calls us into a 'moral community' (Rose) with more-than-human others:

> *III*
>
> [...]
> Pray: pray for Kinglake.
> Pray for the bush and the paddock and the town: pray for the sky and the ground.

> Pray for the possum, the ringtail and brushtail: for the koala and the grey kangaroo.
> Remember to pray for the spider and the skink: the goanna, the gecko, the pink galah.
> Remember the rosella: remember the snake: the heifer, the horse and the brown-speckled hen.
> Remember to pray for Kinglake.
>
> (40)

Echoing the prophetic witness of Jeremiah, Albiston's 'Lamentations' not only perform the work of mourning: they also carry an ominous note of warning. In this way, as well as by inviting the reader to enter imaginatively into the collective grief of those affected by eco-catastrophe, the literature of disaster can help to reframe such calamities as wake-up calls. Disclosing the potential for worse to come and thereby encouraging preventative (which is to say, socio-ecologically transformative) action, literary and other works of the creative imagination, when read in conjunction with other kinds and sources of knowledge, can also engender greater mental preparedness for the shocks that have now become inevitable, fostering endeavours to reduce vulnerability, enhance resilience, seek justice and practise compassion in the increasingly calamitous Anthropocene.

In conclusion, then, I would like to suggest that just as, on this continent, people learnt long ago how to partner with one another and their other-than-human kin in a dance of pyro-symbiogenesis attuned to the requirements of collective flourishing, so now, on this planet, contemporary scholars of human contexts, symbols and meanings are being called into partnership with physical, biological and social scientists, as well as with experts in non-western, and non-modern ways of knowing, in order to improvise the kinds of integrative, transdisciplinary knowledge that are required to help us to dance as best we can with the ever more frequent surprises arising from our increasingly anthropogenic earth, seas and skies.

Acknowledgements

I am grateful for the support of the Humboldt Foundation, together with the Freiburg Institute of Institute of Advanced Studies, where this chapter was written.

Note

1 Personal email from the author, 14 May 2014. Extracts from Jordie Albiston's poem 'Lamentations' are reprinted here with the kind permission of the author.

Works cited

Albiston, Jordie. *XIII Poems*. Melbourne: Rabbit Poets Series, 2013. Print.

Barlow, Karen (Reporter). 'Serious storms head south down Australian east coast', 'Lateline' report, Australian Broadcasting Corporation (28 January 2013). Web.

Berthaller, Hannes, Rob Emmett, Adeline Johns-Putra, Agnes Kneitz, Susanna Lidström, Shane McCorristine, Dana Phillips, Isabel Pérez Ramos, Kate Rigby and Libby Robin. 'Mapping Common Ground: Ecocriticism, Environmental History, and the Environmental Humanities', *Environmental Humanities* 5 (2014): 261–271. Web.

Carrigan, Anthony. '(Eco)Catastrophe, Reconstruction, and Representation: Montserrat and the Limits of Sustainability', *New Literatures Review* 47–48 (2011): 111–128.Print.

Castree, Noel, with William M. Adams, John Barry, Daniel Brockington, Bram Büscher, Esteve Corbera, David Demeritt, Rosaleen Duffy, Ulrike Felt, Katja Neves, Peter Newell, Luigi Pellizzoni, Paul Robbins, Kate Rigby, Libby Robin, Deborah Bird Rose, Andrew Ross, David Schlosberg, Sverker Sörlin, Paige West, Mark Whitehead and Brian Wynne. 'Changing the Intellectual Climate', *Nature Climate Change* 4 (2014): 763–768. Web.

Devine, Miranda. 'Green ideas must take blame for deaths', *Sydney Morning Herald* (12 February 2009). Web.

Estok, Simon C. 'Theorising in a Space of Ambivalent Openness: Ecocriticism and Ecophobia', *ISLE* 15.2 (2009): 203–225. Print.

Evans, Richard. *Disasters That Changed Australia*. Carlton: Melbourne University Press, 2009. Print.

Fischer, J., A. D. Manning, W. Steffen, D. B. Rose, K. Daniell, A. Felton, S. Garnett, B. Gilna, R. Heinsohn, D. B. Lindenmayer, B. MacDonald, F. Mills, B. Newell, J. Reid, L. Robin, K. Sherren, and A. Wade. 'Mind the Sustainability Gap', *Trends in Ecology and Evolution*, 22.12 (2007): 621–624. Print.

Griffiths, Tom. '"An unnatural disaster"? Remembering and forgetting bushfire', *History Australia* 6.2 (2009): 35.1–35.7. Web.

——*Forests of Ash: An Environmental History*. Cambridge: Cambridge University Press, 2001. Print.

Hartman, Steven. 'Inscribing Environmental Memory in the Icelandic Sagas', North Atlantic Biocultural Organisation. N.d. Web.

Hoffman, Susanna M. and Anthony Oliver-Smith (eds). *The Angry Earth: Disaster in Anthropological Perspective*. London and New York: Routledge, 1999. Print.

Jones, Rhys. 'Fire-Stick Farming', *Australian Natural History* (September 1969): 224–228. Print.

Latour, Bruno. *We Have Never Been Modern*. Trans. C. Porter. Hemel Hempstead: Harvester Wheatsheaf, 1993. Print.

Lindenmayer, David, Sam Banks, Lachlan McBurney and David Blair. 'After the Fire: Leadbeater's Long Journey', *Ecos* 157 (2010): 1–5. Web.

Lawson, Henry. *Poems*. Preface and Chronology by Colin Roderick. Sydney: John Ferguson, 1979. Print.

Mathews, Freya. 'Fires the Deadly Reality of Climate Change', *The Age* (10 February 2009). Web.

Nixon, Rob. *Slow Violence and the Environmentalism of the Poor*. Cambridge, Mass.: Harvard University Press, 2011. Print.

Paterson, Alistair. 'Enduring Contact: Australian Perspectives in Environmental and Social Change', *Occasion: Interdisciplinary Studies in the Humanities* 5 (March 2013): 1–17. Web.

Pyne, Stephen J. 'Consumed by Fire or Fire: A Prolegomenon to Anthropogenic Fire', *Earth, Air, Fire, Water: Humanistic Studies of the Environment*. Eds Jill Kerr Conway, Kenneth Keniston and Leo Marx. Amherst: University of Massachusetts Press, 1999. 78–101. Print.

——*The Still-Burning Bush*. Melbourne: Scribe, 2006. Print.

——*World Fire: The Culture of Fire on Earth*. Washington: University of Washington Press, 1997. Print.

Rigby, Kate. *Dancing with Disaster: Environmental Histories, Narratives, and Ethics for Perilous Times*. Charlottesville: University of Virginia Press, 2015. Print.

Rose, Deborah Bird. '"Moral Friends" in the Zone of Disaster', *Tamkang Review*, 37.1 (2006): 77–97. Print.

Rose, Deborah Bird and Libby Robin. 'The Ecological Humanities in Action: An Invitation', *Australian Humanities Review* 31–34 (April 2004). Web.

——'A Manifesto for the Ecological Humanities', Fenner School, Australian National University. 2001. Web.

Sandilands, Catriona. 'Queer Life? Ecocriticism After the Fire', *The Oxford Handbook of Ecocriticism*. Ed. Greg Garrard. Oxford: Oxford University Press, 2014. 305–319. Print.

Serres, Michel. *The Natural Contract*. Trans. Elizabeth MacArthur and William Paulson. Ann Arbor: University of Michigan Press, 1995. Print.

Steffen, Will. 'The Angry Summer', Australian Climate Commission Report (February 2013). Web.

Stretton, Leonard. *Report of the High Commission to Inquire into the Causes of and Measures taken to Prevent the Bush Fires of January, 1939, and to Protect Life and Property and the Measures to be Taken to Prevent Bush Fires in Victoria and to Protect Life and Property in the Event of Future Bush Fires*. Melbourne: T. Rider, Acting Government Printer, 1939. Web.

Teague, Bernard (Chair), Ronald McLeod and Susan Pascoe (Commissioners). *2009 Victorian Bushfires Royal Commission Report*, 2010. Web.

15 Placing the nation

Curating *Landmarks* at the National Museum of Australia

Kirsten Wehner

It's early on a Thursday afternoon and I'm loitering in a corner of the National Museum of Australia's Landmarks *gallery. I could be, if anybody asked, en route to a meeting elsewhere in the building, but to tell the truth I've walked down to the exhibition from my office to spend time with the displays. It's restful to stand in the still, slightly dim light and imagine other lives. I stop towards the end of one case and study the objects: a weathered, termite-mined fence post, a pair of opera glasses, a silver box, and a bundle of spiky brown seeds. I think of how it would feel to touch the striated surface of the wooden post, and notice that some of the seeds have slipped off their plinth on to the floor of the case.*

My eyes drift up to the graphic panel on the rear wall and I read about pastoralist Alfred Giles, who in 1879 founded Springvale station on the Katherine River in what is now the Northern Territory. Giles' wife Mary arrived a year later, bringing with her the glasses and box to decorate her new house. A slightly sepia image on the bottom of the panel shows Alfred with his and Mary's children, the smallest girl perched on a pony, in front of their homestead. The image was taken the same year that the Giles family abandoned Springvale, defeated by local Jawoyn and Dagoman peoples who attacked settlers in defence of their lands, termites that consumed anything made of timber, and spear grass seeds that invaded sheep's fleeces, ruining the wool and eventually penetrating the animals' skins. I know the Springvale story well, but it always draws me back. I can never decide who its hero is.

I turn a bit to the left and gaze at another group of things: a hafted stone axe and a second axe head fashioned from a metal rasp, a delicate beaded armband with a design showing 'HTT' and a rusted branding iron finishing in the same letters, but back to front. Along the front of the case, labels describe how Nym, Mangarrayi elder and stockman, and Amy Dirngayg, Mangarrayi woman and cook, made and used these objects on Elsey, a station south-east of Springvale on the Roper River. Founded two years after Springvale, Elsey prospered. Its managers, including Alfred Giles's son Harold, chose country better suited to grazing stock, especially cattle, and they and the local Mangarrayi and Yangman peoples established more peaceable relations.

Nym and Amy Dirngayg, and others of their communities, became indispensable workers on Elsey, finding a way to live on their Country even as they were exploited through, for example, being paid much less than white employees. In the top left hand corner of the case, a large yellow and black sign reflects the warm glow of the lights. I tilt my head up so I can study the central motif of a black cockatoo in flight. An adjacent panel explains how, in 1972, the Mangarrayi people of Elsey established a settlement away from the station homestead, at a place upriver called Jilkminggan. The sign, featuring the cockatoo, a significant Mangarrayi totem, stood in front of the local school.

I hear footsteps and then giggles and glance over my shoulder. Three girls have stopped behind me, arrested in their hurry through Landmarks *by a taxidermy sheep and a large basket filled with frothy white and grey wool. The girls are probably at the Museum on a school excursion. Their uniforms are a bit of a giveaway, plus they're clutching clipboards with pages of questions interspersed with blank black lines for answers. Lots of teachers still create these 'worksheets,' though they're considered a bit old-fashioned now. These pages always annoy me, suggesting that a visit to the Museum is about gleaning little gobbets of information, sort of like using Google but with more walking.*

The girls stand in a line and stare at the sheep. The sheep stares back. The girls look at each other and then wriggle around and squeak a bit. One stretches out a hand to caress the sheep's back and triggers the proximity alarm, just as another puts her fingers into the wool basket and plucks out a few fibres. She slips them into her pocket. My first impulse is to remonstrate with the girls for touching the objects on display, but I don't. I'm a little thrilled by their naughtiness, so I just sort of pretend that I don't work at the Museum and walk slowly away. I wonder whether, in a week or so, the young woman will remember the worksheet or the wool, which by then will have infused her blazer with the faint aroma of lanolin.[1]

Introduction

Opened in 2011, *Landmarks: People and Places across Australia* is a National Museum of Australia 'permanent' exhibition that traces a history of Australia since European colonisation of the continent began in the late eighteenth century.[2] Bringing together over 1,500 artifacts, texts, images and multi-media elements, the gallery explores ten major themes in the country's past, considering how each has unfolded over time in particular locales across the continent (see Figure 15.1). This chapter explores how *Landmarks* emerged through and expresses curatorial efforts to re-shape museum, and particularly cultural history museum, exhibition practices in more Anthropocenic modes. Drawing expressly on my experience as a *Landmarks* senior curator and, from 2005, the project's content director, I reflect here on how the exhibition endeavours to disrupt modernist tropes of the nation-state and national history, working instead towards a place-based history of Australia that narrativises the past as an inter-connected web of diverse, located experiences.[3]

The idea of the Anthropocene draws attention to the ways in which the Earth's accelerating ecological crises are anthropogenic events and should consequently be understood as cultural rather than techno-scientific problems. The concept suggests that if humanity is to go on at all, then it must, as the Australian environmental philosopher Val Plumwood puts it, 'go on in a different mode' (cited in Gibson, Rose and Fincher vi). Humanity – or at least the privileged, Western world – needs new ways of living; ways that re-work modernist frameworks that separate human history from natural history, mind from matter

Figure 15.1 View into the *Landmarks* gallery, 2012. National Museum of Australia. Photo by George Serras.

and cognition from sensing. As Rose elucidates, following Plumwood, the search for these new ways of living can be productively understood as centering on two major tasks: 'resituating the human in ecological terms' and 'the non-human in ethical terms' (3–4).

Landmarks was not developed explicitly in relation to the idea of the Anthropocene. As the gallery took form between 2004 and 2010, humanities scholars in universities moved quickly to take up this 'meta-concept' (Nixon np).[4] Curators, overall, found the idea less immediately useful. The term – 'Anthropocene' – held little currency outside the academy and was consequently difficult to assimilate within museum institutions geared to public as well as specialised conversations. Curators were also, perhaps, rather slow to consider how the idea extended their long-established interest in the nature of human relationships with material worlds. The *Landmarks* team spent relatively little time debating the Anthropocene, but they were nevertheless centrally engaged, as I elucidate here, with Plumwood's call to build new conceptual frameworks constructing people – and more specifically, non-Indigenous Australians – as part of ecological communities that are also constituted by other acting, purposeful, often sentient, non-human beings.

I begin this chapter by sketching out *Landmarks*' genesis, tracing how the gallery's curators framed it initially to negotiate, on the one hand, calls for the Museum to articulate a singular, triumphal national history oriented towards producing a consensual view of Australian identity, and, on the other hand, curatorial commitment to an inclusive and accessible material history that engaged visitors with some of Australian society's most pressing cultural–environmental challenges. I explore how the curators identified the concept of 'place' as a key tool for managing these discursive ambitions, tracing how this framework extended the Museum's established interest in 'people and the environment,' in conversation with the broader field of the Australian environmental humanities, and responded to certain understandings of the communicative potential of the Museum's collections. I note the significance in *Landmarks*' development of the concepts of 'country,' as in, the nation-state, and 'Country,' as in Aboriginal and Torres Strait Islander cultural–ecological relationships to land and waterscapes (see Rose, this volume).

I then turn to focus on how *Landmarks*' treatment of place, including its construction as a particular kind of spatialised experience, can be understood as curating three key Anthropocenic moves. First, I consider how the gallery represents places as contemporary communities constituted through the efforts of generations, thus rupturing discursive divisions between the past and the present. Second, I explore how the gallery expands the field of historical actors to include not only diverse peoples, Indigenous and non-Indigenous, but also non-human species, natural forces and 'inanimate' technologies. Finally, I reflect on *Landmarks* as, in and of itself, a kind of place in which practices of assemblage, styles of interpretation and spatial choices propose new modes of subjectivity centred on embodied attention and creative response to the material world. I conclude by suggesting that *Landmarks* invites visitors to relate to communities of represented others – Indigenous and settler peoples, termites and wood posts, spear grass and sheep – through creative, embodied, inter-subjective modes of meaning-making that hint productively at new modes of Anthropocene being.

Nation, theme and place

The National Museum was established in 1980 to collect, research and interpret material relating to 'Australian history'. In 2001 the institution opened its major public programmes facility in Canberra, Australia's capital city. Less than two years later, the Museum's Council, its governing body, commissioned an independent review of the institution's exhibitions, and then endorsed the panel's findings. The review argued that the Museum's approach was overly fragmented, indeed incoherent (Commonwealth of Australia), and Council subsequently asked the Museum to develop, essentially, a singular, celebratory national history.

Landmarks was initiated as one of the Museum's key responses to the 2003 review. The curatorial team, then led by Mathew Trinca, was obliged to address Council's expectations. The curators were also committed, however, to continuing Museum traditions of addressing Australia's cultural plurality, including acknowledging violence, dispossession and poverty as part of the national past (Trinca and Wehner). Responding to these tensions, *Landmarks* was framed as a 'general history of Australia's economic, social and political conditions' (National Museum of Australia 1) that would explore 'ten key themes in Australian history,' through 'a focus on events in particular times and places' (6). This approach was pitched as addressing the 'primary narratives of Australia' and significant moments in 'the national story,' while avoiding 'generalising historical conditions across the country' or answering 'abstract questions of national identity' (2).

As *Landmarks* evolved, the gallery's interpretive structure consolidated around ten themes, mapped on the gallery floor to ten 'modules,' or groups of exhibits. Each theme is conceptualised as focusing on a 'grand idea' – an ambition, philosophy or problem – that has impelled people's approaches to living on the Australian continent. The themes are consequently conventional, replicating at a gross level well-known, arguably conservative, conceptualisations of the major drivers of the development of the Australian nation-state. The thematic modules are located in the gallery to suggest a rough chronology, in some ways echoing a developmental national narrative. The gallery begins, in terms of the primary circulation route, with imperial British ambitions to establish new outposts of European society on the Australian continent, then European-Australians' journeys of exploration, the expansion of pastoralism, gold rushes and the growth of democratic movements, the development of agriculture, endeavours in research, scholarship and education, the growth of communication and transport networks, the post-war evolution of the mining, manufacturing and finance industries, and ideals of urban life.[4]

However, while *Landmarks* acknowledges the role of universalising, or nationalising, ambitions in shaping historical experience in Australia, it simultaneously undermines assumptions about their utility in understanding historical experience. Every *Landmarks* module is comprised of three or four 'place' exhibits, each of which explores the history of a particular location (through the perspective of the relevant theme). Theme and place are thus interrelated to accentuate the abstract character of modernist discourses about building Australian society, inviting attention to how people prosecuting these ambitions have everywhere been obliged to re-think their ideas and actions in response to local cultural–ecological conditions.

Indeed, interpretive and design choices throughout the gallery valourise place, rather than theme, as the primary context for making sense of historical experience and, by extension, contemporary meaning, attachment and action. *Landmarks* begins with a large, slow-moving audio-visual 'tour' of the places – their landscapes, people, plants, animals and atmospheres – featured in the gallery, as they appeared in 2010. Each module is introduced by a plain graphic panel presenting a modest one paragraph summary of the theme and noting the relevant place exhibits. Place exhibits, in contrast, begin with colourful, backlit 'markers' featuring a large photograph of the relevant location, a longer textual local history, and physical, often touchable, materials from the place. Exhibit display content, as I elucidate below, is resolutely focused on local histories.

People, environment and objects

The *Landmarks* curators' conceptualisation of place in the gallery drew on and sought to extend the Museum's established traditions in exploring histories of people's interaction with environments, as these are recorded in material culture. When the Museum was established in the early 1980s, 'people and the environment' was constituted as one of the institution's three, inter-related themes (the others being Aboriginal and Torres Strait Islander cultures and histories and Australian history and society since 1788). Initially inspired by approaches in environmental archaeology and human ecology, through the 1990s and early 2000s the people and the environment theme was strongly informed by concepts of 'cultural landscape,' often glossed as 'place'. Senior curators and cultural geographers Richard Baker and Ruth Lane, and later archaeologist Mike Smith and historian Matthew Higgins, led a range of projects exploring how objects expressed diverse settler and Indigenous perceptions and understandings of environments and their associated land uses. Exhibitions were often oriented to developing people's capacities to read landscape histories (Baker and Lane; Smith; Robin). In keeping with the Museum's broader institutional self-definition around 'social history,' these projects tended to centre on how people's attitudes, choices and lives shaped and were shaped and were, particularly, enabled, by non-human species and forces.

The Museum's People and the Environment theme developed in conversation and collaboration with Australian environmental history, and particularly scholars such as Tom Griffiths, Peter Read and Mark McKenna, who were elaborating locale as a key framework for interrogating Australian history, in part because of the way it enabled detailed exploration of Indigenous and settler relationships (see Griffiths). As the *Landmarks* development got underway, the curators also increasingly engaged with the work of scholars such as Libby Robin, Deborah Bird Rose and Kate Rigby, who were then evolving new interdisciplinary, multi-species, ecological approaches to studying human/non-human communities (see Rose et al.). Curator and environmental historian George Main, in particular, elaborated these approaches within the *Landmarks* team, fostering consideration of how collections could be understood as encoding ecological forces (see Main). My own interest in place-based ethnographic research and senior curator Martha Sear's and curator Jennifer Wilson's experience in curating regional and community museums, including those focused on primary industries centred on human–animal relationships, also developed the team's understanding of place as a conjunction of diverse, located agencies, forces, peoples and trajectories (see Sear).

Reflecting these threads, the *Landmarks* team envisaged the gallery as treating place as a means to encompass and bring into relationship diverse Indigenous and non-Indigenous modes of perceiving, using and relating to the land, as well as to explore how non-human species and forces had stimulated, enabled and been shaped through these modes. They also aimed, however, to extend the Museum's existing practice by shifting *Landmarks*' primary interpretive focus away from 'people' and on to 'places' themselves. In a sense, the curators aimed to move from focusing on 'people and the environment' to consider more fully 'environments, and their peoples and non-humans'. This move was oriented, in part, towards re-positioning human agency as one among many different kinds of agency, thereby opening up space to conceptualise non-human agency in shaping places and to represent non-humans as 'beings' travelling their own trajectories, even as these intersected with human lives in the active production of places. Shifting the representational focus on to 'place' was also seen as a way to move away from talking generically about Australian environmental conditions (and particularly their differences from European conditions) to focus instead on how ecological forces acted in particular locales to shape human and non-human lives (even where these forces were poorly recognised in human ideas about places). The curators argued that by focusing on how specific places had evolved over time the gallery would move beyond simply juxtaposing Indigenous and non-Indigenous perspectives on environments to attend to how specific Indigenous and non-Indigenous people, and non-human species and forces, had interacted to produce life-worlds.

The *Landmarks* curators' conceptualisation of place emerged as they brought their reflections on the Museum's established practices in interpreting 'people and the environment' and on environmental humanities scholarship into conversation with their developing sense of the communicative potentials of the Museum's collections. From the early 1980s, the Museum's collecting culture developed to privilege 'secular relics,' that is, 'objects deemed significant by virtue of their association with a particular person, community or event'. Curators were encouraged to collect artifacts with a 'strong 'personal story',' meaning provenance linking the object to an identifiable individual and his or her life story (Wehner and Sear 145), and the Museum developed with a focus on documenting the lives of 'everyday Australians' and, through them, broad trajectories in social and environmental change.

Responding to this collecting history and in concert with contemporary scholarship in material culture studies (see for example, Miller; Dudley; Tilley et al.), the *Landmarks* curators understood objects as embodying aspects of the cultural–material conditions of existence in *particular* times and places, and further the human and non-human interactions that had produced them. The curators argued that the Museum's collections would consequently engage visitors most powerfully when contextualised and interpreted in terms of those conditions and interactions (see Wehner and Sear 153). Drawing on this understanding, the curators envisaged *Landmarks* as bringing together sets of objects that had authentic and organic histories in particular places – that had been made or used in those places and that through their materials, forms and traces of wear and weathering expressed aspects of those histories. Objects displayed in the context of the places in which they had 'lived' were envisioned as enabling visitors to engage imaginatively and empathetically with the lives of the represented peoples and non-humans, opening out embodied explorations of the significance of local cultures and ecologies in shaping others' and, by extension, visitors' lives.

Negotiating 'Country'

The Aboriginal and Torres Strait Islander concept of 'Country' became a key provocation in curatorial resolution of the gallery's interpretive aims, polemical import and emotional tenor (see also Hogan, this volume; Rose, this volume). *Landmarks* was developed initially as *Creating a Country*, the phrase emerging to gloss Council's call for a narrative of the unfolding of the Australian nation-state. The curators rejected the title's teleological nationalism, and the way that 'country' reinforced the frame of the nation, but they apprehended the word – 'Country' – as working complexly in relation to and potentially against nationalist abstractions. 'Country,' capitalised, asserted Aboriginal and Torres Strait Islander peoples' pre-eminence and their ownership of, rights over and in and belonging to lands. The concept grounded and directed attention towards histories of violence, dispossession and negotiation between Indigenous and settler Australians, thus contesting national narratives fantasising about the peaceful evolution of an Australian democracy. More broadly, 'Country,' understood as the integrated totality of a locale's land and waterscapes, its people, plants and animals, past and present, its stories, meanings and practices of custodianship and care, inspired and resonated deeply with the curators' visioning of places as key sites of Anthropocene action.

Country, and its significance in Australian history, is consequently signalled in the gallery in a variety of ways. Almost every exhibit begins, and often ends, with objects revealing Aboriginal histories in places, and particularly practices of living with and caring for environments. The word repeats through the gallery's interpretive text. Graphic elements introducing each place exhibit, for example, begin by identifying the locale as Jawoyn, or Wiradjuri or Birpai country (though in lower case, according to Museum style) and narrate Aboriginal experiences of colonisation in that place. In gallery modules like 'Exploring the Country,' the word is used more ambiguously. It is linked, in this instance, to settler Australian journeys of exploration that, while understood by European-Australians as heroically opening the continent for colonisation and ultimately nation-making, also unavoidably revealed to them 'Countries' that were already deeply 'explored' and 'mapped' – known by Aboriginal and Torres Strait Islander peoples.

Curatorial engagement with understandings of Country also helped solidify interpretive emphases on the immanence of the past in the present, on human material, spiritual and cultural inter-connectivity with land and water, on multi-species agencies, and on the need for an ethic of care and respect in relation to places. Indeed, at various stages, the curators were so inspired by the concept that they proposed that the gallery be called something like 'Creating Countries'. In the end, however, the use of 'country,' let alone its capitalised or pluralised versions, failed to find a place in the title. Museum stakeholders, including marketing officers and senior Aboriginal staff, argued against its inclusion on the grounds that the public might think the gallery was about Indigenous history and culture. For some, *Landmarks* was seen as 'white history,' with Aboriginal and Torres Strait Islander history constructed as the province of the Museum's *Gallery of First Australians*.

Past and present

Responding to these negotiations, the *Landmarks* curators constructed the gallery to connect past and present, bringing into tension understandings of time that emphasise linearity with others that evoke simultaneity. The gallery as a whole asserts the 'here and now' as the context for understanding represented places, while also offering a series of semi-chronological accounts of places. Each exhibit is introduced as representing a contemporary place, with markers referring to locales by their contemporary name, presenting images of places as they appeared just before the gallery opened and explicitly narrating the relevant place history to the present day. Each exhibit then centres on a bounded space, such as a single large case, displaying a series of 'stories' comprised of objects, images and text elements, each referencing a particular period in a place's history. These stories are arranged to create a selective chronology of the locale stretching from the late eighteenth or early nineteenth century to the early twenty-first century, while also incorporating references to deeper Indigenous and non-human histories.

The Blue Mountains exhibit, for example, part of the 'Exploring the Country' module (see Figure 15.2), begins at one end of the display case with the first successful ascent of the range by settlers Gregory Blaxland, William Wentworth and William Lawson, who in 1813 followed Aboriginal pathways across the mountains to open up inland New South Wales for pastoral settlement. The display continues with transformations of the range through, for example, road building, and concludes with the growing numbers of bushwalkers, conservationists and tourists who have explored the region since the late twentieth century. In the

Figure 15.2 Blue Mountains exhibit, *Landmarks* gallery, 2012. National Museum of Australia. Photo by Jason McCarthy.

centre of the exhibit, a large painting, created in 2008, depicts the view inland as Blaxland, Lawson and Wentworth might have seen it from the top of the range, the blue bush now fetishised in tourist advertising broken with trails of smoke referencing Aboriginal camps.[6]

The curators envisaged the gallery as replacing constructions of the national past as a single, linear unfolding with a model of Australian history as a web of concurrently evolving, linked trajectories in which past and present did not exist in any simple sequential relationship and indeed could not be understood as somehow essentially discontinuous. Objects' capacities to collapse spatio-temporal scales, to be experienced as simultaneously of past and present, were central to this vision. Each exhibit thus envelopes artifacts from diverse historical moments, asserting their co-existence outside any one time, while also relating them through interpretive elements to relevant historical periods and to the contemporary moment. Displayed objects exist as standing for the times and places in which they were made, used and discarded, and as present in the gallery and the lives of its visitors. In a sense, *Landmarks* embodies in its materiality the immanence of the past in the present.

The curators envisaged *Landmark*'s treatment of temporality as inter-weaving past and present in a representational sense, but also as inviting visitors – quite obviously in the present – to consider themselves in relation to and perhaps within the gallery's narratives. Through constructing chronological place histories, the curators sought to develop visitors' capacities to apprehend how places had been historically produced, and were consequently both deserving of respect as the product of past endeavours and malleable and open to re-working. Through establishing the present as the gallery's overarching interpretive framework, introducing places as contemporary locales and extending display chronologies into the twenty-first century, the curators also invited visitors to understand represented places not as historical curiosities but rather, as of the same kind as visitors' own places, and vice versa. The curators sought to engage visitors with how people in many ways like themselves were part of the on-going stories of place, inheritors of pasts and responsible for how contemporary choices and actions might play out in the future. Indeed, the curators endeavoured to suggest to visitors that they, as much as any of the people constructed as national heroes, such as Blaxland, Lawson and Wentworth, were, and could be, participants in constructing future worlds.

Multi-species places

The *Landmarks* curators aimed to develop visitor attention to and understanding of peoples,' both Indigenous and non-Indigenous, agency in shaping places. They also sought to reposition human agency within a broader field of place-making actors. In developing each exhibit, the curators selected objects that, individually or as a group, related to three, inter-connected threads: natural forces and non-human species, diverse Indigenous lives and experiences, and similarly differentiated settler lives and experiences. In a sense, the curators aimed to expand the field of 'players' shaping historical trajectories, including not just diverse peoples but also rocks and rivers, kangaroos and cattle.

The exhibits were developed as complex, dense, overlapping object assemblages that bring into close conversation many different kinds of objects: taxidermy animals and artworks, documents and throwing sticks, stone tools and death masks. Interpretive text elements, and particularly object labels, narrate each artifact's specific history, teasing out how it embodies local materials and modes of living and bears the traces of conditions and interactions that have shaped it over time. Displayed objects begin, through their materials, forms and finishes, to express the distinctive physicality and cultural history of each place. As visitors bring their own bodies into conversation with the exhibit, examining, responding to and connecting up different elements, the displays perform the inter-twining interactions between different agents through which places were and continue to be produced.

As an example of such an assemblage, the Hobart exhibit of the 'Colonial Foundations' module begins with a large taxidermy Forester kangaroo, a sub-species of the Eastern Grey found in Tasmania, Aboriginal stone cutting and scraping tools from the Derwent River area and the diary and death mask of the Reverend Knopwood, one of the first settlers in the colony (see Figure 15.3). Interpretive elements elucidate how the local Mouheneenner and Moomairremener peoples managed grasslands in the region to increase numbers of kangaroo, as well as other game, and how, as Knopwood describes in his diary,

Figure 15.3 Hobart exhibit, *Landmarks* gallery, 2012. National Museum of Australia. Photo by Jason McCarthy.

British colonists also soon came to rely on the animals for food. As convict shepherds moved inland, they adopted practices such as wearing kangaroo skin clothing based on local Aboriginal life-ways, enabling them to live off the land far from colonial authority, while Aboriginal peoples incorporated British dogs into hunting practices. Kangaroo numbers declined, exacerbating tension between Indigenous and settler peoples.

Following groups of objects trace out these inter-weaving trajectories. A prisoner's punishment shoe, teamed with a lump of quarried rock, evokes the highly punitive system of law and order introduced in response, in part, to convict capacity to survive beyond settlements. Prosperous pastoralists' domestic goods are grouped with a 'proclamation board' created by the colonial governor to assert the equality of all peoples under British law just as he moved to expel Aboriginal people from the 'settled' areas of the colony. The exhibit concludes with a kelp water carrier made in 2009 by artist Leonie Dickson, descendant of the Tasmanian Aboriginal community that survived colonisation and now continue to work with the distinctive plants and animals of their Country.[7]

Constituting visitors

Detailing their manifesto for living in the Anthropocene, Anderson et al. argue that 'we' need new 'unconstrained and expansive' ways of thinking about and relating to the world. They suggest that:

> This means … listening to the world. This means giving up delusions of mastery and control, and instead seeing the world as uncertain and yet unfolding. So our thinking needs to be curious, experimental, open, adaptive, imaginative, responsive and responsible.
>
> (i–ii)

In other words, these authors call for the development of new kinds of human subjectivity, ones centred on attending closely to the diverse human and non-human communities within which we are enmeshed, and on responding and adapting to those communities not in terms of seeking to penetrate, dominate and establish their essential truths, but rather in curious, creative and open-ended ways.

In concert with this argument, the *Landmarks* curators envisaged the gallery as embodying design choices that invite visitors to modes of 'interactive dwelling,' centered on active, multi-sensorial, open-ended engagements with object displays. Across the gallery, each thematic module is constructed as a porous 'enclosure,' with a large central plinth presenting objects on open display flanked by cases – one or two to each place exhibit – displaying dense arrangements of smaller objects (see Figure 15.4). Visitors enter and are encouraged to pause within each module, sometimes through the creation of a circulation cul-de-sac. They need to shift their gaze around and across various exhibit elements, turn their bodies and move positions in order to develop a sense of the whole

Figure 15.4 View into the 'Gold and Government' module of *Landmarks*, showing the Bendigo exhibit case and central plinth and the Lachlan Valley exhibit to the rear. A gold cradle interactive sits in the back right-hand corner. National Museum of Australia. Photo by Jason McCarthy.

environment and to inspect elements within it. In a sense, the design invites viewing 'in the round,' locating visitors within the represented world and subverting attempts to take up a position outside it. At the same time, the design resists valourising any singular route through the gallery, as multiple paths around plinths and views through cases and walls to different parts of the gallery invite visitors to shape their own narrative journeys through the exhibition.

Landmarks is object rich, tuning visitors to the significance of the material world by quite simply enabling objects of varying sizes, shapes and materials to dominate the gallery's visual field. Within display cases, objects are positioned at the centre of the visitor's gaze and grouped in ways that encourage visitors to track visually back and forth between artifacts, without over-determining which objects the visitor will thereby connect up into narrative threads. Graphic panels carrying interpretive text are positioned up and to the back of objects, presented in regular sequences that reduce their tendency to attract the eye. Object labels are pulled to the front of cases and located below primary sight lines. Visitors are thereby invited to articulate their bodies to the objects on display, to attend to them as the primary expressive elements within the exhibition. This suggestion is reinforced by the interpretive text focusing on how particular objects were produced through and participated in people's and non-humans' historical experiences. *Landmarks* text labels do not explain what an object 'means,' or define why the curators found it significant, but rather concentrate on providing

Figure 15.5 Visitors engage with a 15.2 tonne excavator bucket from Mount Tom Price iron ore mine in the 'Expanding the Economy' module, *Landmarks* gallery, 2012. National Museum of Australia. Photo by George Serras.

particular historical contexts that invite visitors to imagine others' lives in particular times and places.

Other gallery elements are similarly intended to tune visitors to embodied responses to the material on display and to how these produce a particular kind of knowledge about and attitude of connection with the world (see Figure 15.5). Visitors are encouraged to recognise experiences of scale by positioning themselves in ways that articulate their bodies to comparatively larger or smaller objects, or to both simultaneously. Taxidermy animals are often positioned to meet visitors face on as they enter a space or to hold visitors' gazes as they turn around a corner. Tactile elements are included in many exhibits, often reflecting curatorial anticipation of visitors' desire to touch fragile objects. In the Derby exhibit, for example, a taxidermy steer pokes its head over a barrier carrying a section of hide from one of its herd. As visitors look at the steer, their fingers caress the hair and skin. The Bendigo exhibit invites visitors to rock a replica nineteenth century mining cradle, listening to the rush of pebbles and water that characterised the region's gold fields. These 'sensory stations' are interpretively open-ended, aimed at stimulating haptic responses and sensorial reflections on the places represented in the gallery.

Conclusion

The National Museum of Australia's *Landmarks: People and Places across Australia* gallery aims to develop a more Anthropocenic exhibitionary practice, within the context of the Museum's evolving institutional trajectory and conceptualisation of Australian history. The gallery evolved as something of a compromise between the Museum Council's investment in a singular, celebratory national history and the institution's established curatorial understanding of the inadequacy of 'the nation' as a frame for understanding Australian history. The *Landmarks* team negotiated these tensions by developing a gallery structured around thematically linked representations of places. *Landmarks*' acknowledges the roles of universalising ideas – such as empire building, pastoral expansion and the search for an ideal urban design – in shaping historical experience, but it also problematises these themes as abstract ambitions by exploring how people pursuing them have been transformed by local cultural–ecological conditions, even as they have also transformed those conditions.

While *Landmarks* retains a strong engagement with the concept of 'the nation,' not least by virtue of its location within and as part of the National Museum of Australia, I suggest here that the gallery also goes some way to disrupting universalising and abstracting modes of history. It invites visitors to connect in embodied, imaginative and empathetic modes to the particular material conditions and lived experiences of places. Exhibits argue that diverse human and non-human actors and forces interact to produce places, and that these choices, actions and interactions persist in shaping places in the present. The gallery asks visitors to understand themselves as participants in the on-going stories of places, to attend deeply, dynamically and creatively to the realities of the material worlds in which they live, and, as such, to become more able to, with respect and care, re-shape those places for generations to come. As such, *Landmarks* suggests intriguing new directions for a more Anthropocenic museology.

It's a few months later and I'm sitting in a meeting in the Museum offices, feeling guilty about the taxidermy sheep. The last few days have been difficult. After presenting a talk about Landmarks *at a conference, I was grilled on the politics and ethics of displaying animal bodies. 'Museums don't need to just keep doing the same thing,' I was told. 'They don't need to keep reproducing human mastery by exhibiting animals like objects.' I got flustered and answered poorly, in part because the comment implied good questions. Does taxidermy invite people's awareness of non-humans as living beings? How can artifacts speak to animal agency or sentience? Then, yesterday, I spent an hour or so talking with a visitor who was complaining that we had chosen to exhibit a variety of sheep other than the iconic 'fine-wool merino who built Australia'.* Landmarks *displays an individual from a dual-purpose breed developed to suit local conditions at Springfield station, near Goulburn, New South Wales and to help Springfield's owner ride out market fluctuations by proving valuable for both wool and meat. Now we're discussing how visitor caresses, after depleting the wool basket, are dirtying and possibly damaging the taxidermy specimen. I smell lanolin in my mind's nose, and remember that the journey of transforming our ways of life in the Anthropocene is one of a thousand cuts, clips, er, negotiations.*

Acknowledgements

I gratefully acknowledge the Rachel Carson Center for Environment and Society, Munich, Germany, for enabling me to prepare this chapter through a 2015–2016 Carson fellowship, and the National Museum of Australia for supporting my sabbatical in Munich. I am deeply indebted to my curatorial colleagues with whom I shared the *Landmarks* adventure, and particularly Martha Sear, Daniel Oakman, George Main, Jennifer Wilson and Mathew Trinca. Their creativity, cleverness and collegiality inspire, inform and are discussed in this chapter. The views expressed here, however, are my own and do not necessarily represent those of the National Museum of Australia or my colleagues.

Notes

1 *Landmarks* Springvale and Elsey exhibit was developed by George Main.
2 'Landmarks' Hobart exhibition was developed by Kirsti Graham, Anne-Marie Condé, and Martha Sear.
3 In this chapter, I refer to the *Landmarks* curatorial team in the third person in recognition of the way in which the gallery was, in every way, the outcome of a collaborative, collegial effort. I believe that the team reached a strongly shared vision for the gallery, but my discussion here reflects my own perspective and understanding of the process of producing the exhibition.
4 See Humanities for the Environment (http://hfe-observatories.org/common-threads/) for an overview of the development and aims of the 'Anthropocene Humanities' and for links to relevant scholars and projects across the globe.
5 Oakman, Sear and Wehner present a book-form history of Australia reflecting *Landmarks*' framework and including some of the gallery's stories and objects. See also: www.nma.gov.au/exhibitions/landmarks/home
6 'Landmarks' Blue Mountains exhibit was developed by Pip McNaught, with additional curatorial development by Matthew Higgins and Anne-Marie Condé.
7 'Landmarks' Hobart exhibition was developed by Kirsti Graham, Anne-Marie Condé, and Martha Sear.'

Works cited

Anderson, Kay et al. 'Manifesto for living in the Anthropocene', *Manifesto for Living in the Anthropocene*. Eds Katherine Gibson, Deborah Bird Rose and Ruth Fincher, Brooklyn, NY: Punctum Books, 2015. i–iii. Print.

Baker, Richard and Ruth Lane. *A Changing People – A Changing Land: Options for an exhibition by the National Museum of Australia to tour the Murray-Darling Basin*. Unpublished exhibition prospectus. Canberra: National Museum of Australia. 1992? Print.

Commonwealth of Australia. *Review of the National Museum of Australia: Its Exhibitions and Public Programs*. Canberra: Department of Communications, Information Technology and the Arts, 2003. Print.

Dudley, Sandra, ed. *Museum Materialities: Objects, Engagements, Interpretations*. London and New York: Routledge. 2010. Print.

Gibson, Katherine, Deborah Bird Rose and Ruth Fincher. 'Preface', *Manifesto for Living in the Anthropocene*. Eds Katherine Gibson, Deborah Bird Rose and Ruth Fincher. Brooklyn, NY: Punctum Books, 2015. v–viii.

Griffiths, Tom. 'Environmental History, Australian Style', *Australian Historical Studies* 46.2 (2015): 157–173. Web.

Main, George. 'The Waterhole Project: Locating Resilience', *Manifesto for Living in the Anthropocene*. Eds Katherine Gibson, Deborah Bird Rose and Ruth Fincher. Brooklyn, NY: Punctum Books, 2015. 63–69. Print.

Miller, Daniel. *Stuff*. Cambridge: Polity Press, 2009. Print.

National Museum of Australia. 'Australian Journeys and Creating a Country'. Unpublished exhibition brief. Canberra: National Museum of Australia, 2004. Web.

Nixon, Rob. 'The Anthropocene: The Promise and Pitfalls of an Epochal Idea', *Edge Effects*, 6 November 2014. Web.

Oakman, Daniel, Martha Sear and Kirsten Wehner, eds. *Landmarks: A History of Australia in 33 Places*. Canberra: National Museum of Australia, 2013. Print.

Robin, Libby. 'The Love-Hate Relationship with Land in Australia: Presenting "Exploitation and Sustainability" in Museums', *Nova Acta Leopoldina* 114.390 (2013): 47–63. Print.

Rose, Deborah Bird. 'The Ecological Humanities', *Manifesto for Living in the Anthropocene*. Eds Katherine Gibson, Deborah Bird Rose and Ruth Fincher. Brooklyn, NY: Punctum Books, 2015. 1–6. Print.

Rose, Deborah Bird, Thom van Dooren, Matthew Chrulew, Stuart Cooke, Matthew Kearnes and Emily O'Gorman. 'Thinking through the Environment, Unsettling the Humanities', *Environmental Humanities* 1 (2012): 1–5. Web.

Sear, Martha. 'History in Communities', *Australian History Now*. Eds Anna Clark and Paul Ashton. Sydney: University of New South Wales Press, 198–214. Print

Smith, Mike. 'A History of Ways of Seeing the Land: Environmental History at the National Museum of Australia', *Curator: The Museum Journal* 46.1 (2003): 7–14. Web.

Tilley, Christopher, Webb Keane, Susanne Kuechler-Fogden, Mike Rowlands and Patricia Spyer, eds. Handbook of Material Culture. London: Sage Publications, 2006. Print.

Trinca, Mathew and Kirsten Wehner. 'Pluralism and Exhibition Practice at the National Museum of Australia', *South Pacific Museums: Experiments in Culture*. EdsChris Healy and Andrea Witcomb. Melbourne: Monash University ePress, 2006. 6.1–6.14. Web.

Wehner, Kirsten and Martha Sear. 'Engaging the Material World: Object Knowledge and "*Australian Journeys*"', *Museum Materialities: Objects, Engagements, Interpretations*. Ed. Sandra Dudley. London and New York: Routledge, 2010. 143–161. Print.

16 The oceanic turn

Submarine futures of the Anthropocene

Elizabeth DeLoughrey

> The living ocean drives planetary chemistry, governs climate and weather, and otherwise provides the cornerstone of the life-support system for all creatures on our planet […] If it dies, we die. Our future and the state of the oceans are one.
>
> Sylvia Earle

The oceanic turn

The rise of environmental scholarship in recent decades, including the fields of ecocriticism and the environmental humanities, has been simultaneous with an 'oceanic turn'. Generally speaking, the former fields represent interdisciplinary engagements with the representations of territorially-based place and have recently begun to theorise mobility and displacement across biotic, regional, national, and (post)colonial boundaries. In contradistinction, the field of what I call 'critical ocean studies' has shifted from a long-term concern with mobility and fluidity across transoceanic surfaces to theorising ways of embedding, animating, and submerging, rendering vast oceanic space into place. This turn to ontologies of the sea and its implications for temporality and aesthetics in the Anthropocene will be the focus of this chapter.

The oceanic turn of the twentieth century arose from a handful of developments in both geopolitics as well as disciplinary regroupings in the humanities and social sciences. These discourses are entangled with earlier colonial doctrines of *terra nullius* or nobody's land, better defined here as *aqua nullius*. The oceanic turn can be traced to the 1945 Truman Proclamation –the most significant – and yet largely unremarked – twentieth century remapping of the globe, which extended US territory out to 200 nautical miles known as an Exclusive Economic Zone (EEZ).[1] Due to Truman's annexation of Micronesia, this new EEZ was enormous, tripling the territorial size of the United States. This in turn created a scramble for the oceans, catalysing 200-mile EEZ declarations by nations all over the world and a UN *Convention on the Law of the Sea* that reterritorialised 70 per cent of the planet. Although unnoticed by a rising field called the 'Blue Humanities' (Mentz), Cold War geopolitics had a decisive influence in remapping our understanding of the terraqueous globe.

The second influence on the rise of critical ocean studies is the post-1970s 'spatial turn' in geography, which led to the emergence of globalisation and diaspora studies. Building on the work of LeFevre, Foucault and others, Marxist geography was integral to defining the post-Fordist era of global capitalism through the lens of what David Harvey has called 'historical geographical materialism' (8). This loosening of nationally-bounded modes of thinking about capital and space, coupled with a rise of migration and new formulations of ethnicity led to an unprecedented number of transoceanic studies, notably the work of Marcus Rediker which helped inspire Paul Gilroy's *The Black Atlantic*, a text that inaugurated a new generation of thinking about race in transoceanic ways. This in turn was adapted by scholars theorising the Caribbean as well as the Pacific and Indian Oceans.[2] While the scramble for the seas largely figured the ocean and its resources as subject to the exploitation of discrete national territories, a kind of *aqua extractio*, the work of geographers, historians, and cultural studies scholars configured the ocean as a historical space of transnational capital, empire, slavery – often based on an unmarked masculinity that we might term *aqua homo*.

Third, the spatial turn away from the nation state towards the seas was influenced by the post-independence melancholia of the 1990s, where disappointments with the postcolonial state as a structure for governing the human subject led to a turn to the ocean as a site of 'flows' and 'fluidity,' seemingly outside the territorial and legislative limitations of the state (see Benítez-Rojo 1992). Following Gilroy and others, the ethnically exclusive and hierarchical model of national belonging might be imaginatively transcended by turning to those spaces of fluidity and creolisation. As such, the ocean became a space for theorising the materiality of history, yet it rarely figured as a material in itself (DeLoughrey *Routes and Roots* 22; Steinberg *The Social Construction of the Ocean* 245).With some exceptions, these narratives largely represent a *transoceanic* imaginary, positioning the sea as a stage for human history; a narrative of flat surfaces rather than immersions.

This chapter examines factors that have contributed to the rise of a new oceanic imaginary for the twenty-first century: the global sea level rise that is our visible sign of climate change; the emergence of the environmental humanities, especially scholarship in multispecies ethnographies; and finally the work of Caribbean writers and artists who have long theorised the ocean in terms of the violent convergence of environment and history.[3] In the Caribbean the ocean has long been understood as a material entity; it is an ecology for 'subtle and submarine' poetics in the words of Derek Walcott ('The Sea is History' 138). For Martinican theorist Édouard Glissant, the sea signifies an abyss, a repository for the untold histories of the middle passage. These particular Caribbean writers have imagined a submarine temporality in which linear models of time are distorted and ruptured. This engagement with temporality is the product of the violence of transoceanic colonial history as well as immersion in the materiality of the ocean itself. This question of temporalising the sea is vital because unlike terrestrial space – where one might memorialise and narrativise a space into place

– the perpetual circulation of ocean currents means that the sea dissolves phenomenological experience, and defracts the accumulation of narrative. As we look towards an oceanic future caused by sea level rise, these questions of temporality become vital to understanding the epoch termed the Anthropocene. The challenges about representing the more-than-human temporalities of the ocean, what I'm theorising here as 'sea ontologies,' are addressed by British artist Jason deCaires Taylor, whose submarine Caribbean sculptures he describes as 'moments in passing' (*The Underwater Museum* 8), subject to the erosion and transformation by maritime currents as well through multispecies relations with fish, algae, sponges, hydrozoans and coral.

Oceanic futures

The ocean drives our global climate, and our planetary future is becoming more oceanic. Scientific discourse has positioned the ocean as an evolutionary origin for life on earth and, given the imminent threat of sea level rising, our anticipated destiny. Sea level rise is perhaps our greatest sign of planetary change, connecting the activity of the earth's poles with the rest of the terrestrial world, producing a new sense of planetary scale and interconnectedness through the rising of a world ocean. The emergence of the environmental humanities as a discipline at the start of this century is coterminous with the turn to the Anthropocene, a term some scientists are using to describe an epoch in which the scale of anthropogenic carbon emissions positions humanity as a geological force. This new understanding of the human in relationship to a warming planet has catalysed vital questions about positioning the subject in the discourse of species, history, environment, and politics. After decades of work that examined the historicism and difference of the human subject, particularly in postcolonial studies, we are seeing a discursive shift to figuring humanity on a planetary scale, as Dipesh Chakrabarty has argued. This chapter builds on this work but troubles it by engaging interspecies ontologies in an era of sea level rise that is shaping new oceanic imaginaries.

Christopher Connery has argued that in western discourse, the sea 'emerged as a kind of sublime and inassimilable other, not so much a power over humans, but an element outside and alien to the human, and outside time' ('There Was No More Sea' 502). Yet the Anthropocene has catalysed a new oceanic imaginary in which, due to sea level rise, the largest space on earth is suddenly not so external and 'alien' to human experience. The increase in extreme weather events are correlated to the cinematic visuality of flooding and tsunamis, in which footage of a king tide in Tuvalu for instance comes to stand in for the world's rising ocean. This new oceanic imaginary has inspired an increase in a body of literature, arts, film, and scholarship concerned with our watery futures. In popular U.S. culture, there are sensationalist accounts of an active, threatening ocean in films like Roland Emmerich's *2012* as well as in books like Brian Fagan's *The Attacking Ocean*. The twenty-first century has also seen a rise in documentaries that figure the ocean as a threatening agent against the Pacific Islands in *Rising Waters*

(Torrice) and *Time and Tide* (Bayer and Salzman). A body of work argues for the importance of visualising climate change through film and photography, bringing images of the poles and tropics to a global audience in order to encourage social and political change.

There are geopolitical, biopolitical, environmental and ontological dimensions to this oceanic turn. Some texts figure the ocean as a space for evolutionary, religious and ontological origins and destiny. These tend to emphasise the radical interiority of the sea to the human species. For instance, Jacques Cousteau explains 'our flesh is composed of myriads of cells, each one of which contains a miniature ocean ... comprising all the salts of the sea, probably the built-in heritage of our distant ancestry' ('The Perils and Potentials of a Watery Planet' 13). Elisabeth Mann Borgese, one of the founding members of the Club of Rome as well as the first Convention on the Sea (1970) writes that 'every human [...] is a good bit of planet ocean: 71 per cent of his substance consists of salty water, just as 71 per cent of the earth is covered by the oceans' (*The Oceanic Circle* 258). Other narratives are less naturalising, documenting a new era of empire and territorialism evident in the collapse of fisheries, the transnational corporate scramble for minerals and microbiota in the thermal vents of the Pacific Ocean, and the competing state territorialisms now visible in the Arctic as the ice begins to melt. Whether one considers the seabed, the creatures of the sea, or its surface, the ocean has become a new frontier for capital evident in the libertarian Seasteading Institute that seeks to establish a free state on the high seas and, as their website claims, to 'open humanity's next frontier' (Seasteading website; see also Steinberg, Nyam and Caraccioli).[4] This simultaneous rendering of the sea as open frontier and endless natural resource Helmreich refers to as 'blue-green capitalism' (*Alien Ocean* 26). The general lack of attention to these territorial developments might be attributed to the ocean's figuring as 'capital's favored myth-element' (Connery 'The Oceanic Feeling' 289), creating a lacuna precisely where we should be able to trace the intersections of capital and empire, as well as their impacts on human and nonhuman sovereignty.

Whether it is a figure for the salinity of human blood, a utopian space of biocapital, or the dystopian futurity of climate change, the ocean is represented in an 'oscillation between [...] familiar and strange, as us and not us' (Helmreich *Alien Ocean* 403). It is this figuring of the ocean as the uncanny, the *unheimlich*, and borrowing from Gayatri Chakravorty Spivak, as a figure for 'planetarity' that I explore in the remainder of this essay.

Sea ontologies

Caribbean aesthetics articulate what I am calling 'sea ontologies,' a concept that builds on Philip Steinberg and others who seek to 'develop an epistemology that views the ocean as continually being reconstituted by [...] the non-human and the human, the biological and the geophysical, the historic and the contemporary' (Steinberg 'Of Other Seas' 157). While Steinberg has argued persuasively for reading the ocean as a dynamic force rather than place, for 'decentered ontologies of connection' (161), in the Caribbean the enormity of the 400-year long transoceanic history of slavery and indenture has created an aesthetics that imaginatively populate the sea in an act of regional historiography and ancestral memory. The Atlantic in this regard is understood as an unmarked grave site, and memorialising the enormity of loss for the millions who crossed its expanse has particular material challenges, given that first, the mobility of ocean currents means that we cannot localise its waters as a phenomenologically experienced place, and secondly because there are no accurate recordings of where exactly slaves leapt or were thrown overboard. As such, this is an oceanic archive that lacks the place-based narrative and rituals for memorialisation. The earth-based sacralisation of place is generally rendered by the ritualised placement of bodies, bones, and stone monuments. In order to localise an event that can never be truly historically localised, Caribbean writers have peopled the seabed with human bones, imaginatively figured in the limestone structures of coral reefs. Thus what would be archeology in a terrestrial context becomes submarine diving for an oceanic archive, for the remnants of imperial debris and ancestral origins. For instance, the sailor of Walcott's 'The Schooner *Flight*' observes that 'this Caribbean so choke with the dead'. Only by melting in the water can he see:

> … them corals: brain, fire, sea fans,
> dead-men's-fingers, and then, the dead men.
> I saw that the powdery sand was their bones
> ground white from Senegal to San Salvador.
> (*Collected Poems:* 349)

In contrast to the tourist's narratives of azure seas that are subject to the visitor's desires (*aqua nullius*), Walcott's seas are 'choked' with the visible remnants of living history. This is a living graveyard in which one's interpellation of the submarine is interrupted by a rupture in narrative articulation, in breath. The process of anthropomorphising the corals, from brain corals to dead men's fingers, leads to a visual poetics ('I saw them') of the submarine debris of human history. Thus it is by interpellating nonhuman life forms as human bones that enables the visibility or recognition of a submarine human history, that which resides outside (and below) the official archive.

Walcott's poem suggests that 'subtle and submarine' human histories can only be understood in an active engagement – in dissolving in the spaces of alterity. They render not an active recuperation of the ancestral human, but its constitutive remnants, 'powdery sand' that is, like earthly soil, constitutive and grounding, yet also signifying more-than-human history. Laura Ann Stoler has read Walcott's work through the lens of Walter Benjamin, arguing for a reading of ruins as 'petrified life,' traces that mark the fragility of power and the forces of destruction. She focuses on artifacts and the 'dead matter' of imperial history ('Imperial Debris' 196). Yet Walcott's poetics have long engaged *living* matter as a site for more-than-human history, depicting multispecies engagements with plants, fish, corals, and other creatures of the tropical coast as a means of posing an alternative narrative for history making.

Walcott's submerged narrator invokes 'sea ontologies,' a term I'm using to expand on Elizabeth A. Povinelli's theorisation of 'geontologies,' a mutually constitutive relationship between biography/geology, drawn from Indigenous contexts that destabilise western binaries between figures of life and nonlife (Povinelli n. pag.).[5,6] Povinelli focuses on the deep time of Indigenous knowledges in Australia, so the term 'sea ontologies' might characterise the connection between ancestry, history and non-Western knowledge systems in submarine aesthetics. While the focus has tended to be more anthropocentric, in Caribbean literature and cultural studies the ocean has been figured as an evolutionary and cultural origin in the wake of the brutal loss of ancestral memory, a site of memorialisation, excavation, mourning, the sacred, and history. This is why it was not surprising that when Jason deCaires Taylor began sinking life-sized human sculptures under the Caribbean Sea that the majority of viewers assumed it was an act of memorialising the lost lives of the middle passage. But interestingly, the sculptures' temporal and multispecies engagements are more complex, suggesting that the ocean as medium can symbolise the simultaneity or even the collapse of linear time, reflecting lost lives of the past and memorialising – as an act of anticipatory mourning – the multispecies lives of the future.

Vicissitudes

In his imagination of the 'earth after us,' the future of the Anthropocene, Jan Zalasiewicz anticipates that 'our drowned cities [...] would begin to be covered by sand, silt, and mud, and take the first steps towards becoming geology. The process of fossilization will begin' (272). Yet these drowned cities are already in the making, turning the future subjunctive into present participle. With a series of eco-art cement sculptures sunken off the coasts of Mexico and Grenada entitled 'Vicissitudes,' 'The Last Supper,' 'Time Bomb,' 'The Bankers' and 'Anthropocene,' deCaires Taylor has been levying a submarine critique of, to draw from one of his sculptures, the social and political 'Inertia' in response to anthropogenic climate change. 'The Bankers' (2011) show a group of men in suits, who have their heads in the sand.[6] Other sculptures also critique the capitalist consumption that led to the Anthropocene; the two-ton sculpture 'Inertia' (2011) is of an overweight man, sitting on a couch in front of a television with a plate of fast-food (hamburger and fries) on his enormous lap, and a pile of plastic debris at his feet.[7]

The sculptures are largely allegorical and commentaries on the disorienting effects of temporality in the Anthropocene. As deCaires Taylor notes, 'Over the last 20 years our generation has encountered rapid change technologically, culturally and geographically. I feel this has left us with an underlying sense of loss. My work tries to record some of those moments'.[8] Reflecting a planetary 'crisis of futurity' (Pratt 'Planetary Longings' 211), it produces an affect of mourning and, for this particular body of work, stillness amidst the tremendous pressure and mobility of seawater. Connery has written about how the technologies of globalisation helped contribute to 'the dematerialization' of the sea: 'The oceanic [...] is rather the space of pure distance, a meaningless materiality transcended by instantaneous information flow. Connectivity itself functions to dematerialize the connector, the space between' ('Oceanic Feeling' 296). Notwithstanding the fact that those information flows are dependent on tens of thousands of miles of submarine cables, deCaires Taylor's response to the globalising 'disembedding' from time and space (Giddens 188) is to *rematerialise* the ocean and, by life casting local people for submersion, enables sea ontologies, rendering uninhabitable space into anthropomorphised place.

The 'first underwater sculpture park' is in Moilinere Bay, Grenada, established in 2006 by the Grenadian government and tourist board including sixty-five sculptures that have been called by National Geographic as one of the 'Wonders of the World'. The second, the Museo Subaquático de Arte (MUSA), in The National Marine Park of Cancun, Mexico, was established in 2009. At both sites deCaires Taylor works with local artists, students and ecologists, spends months making plaster and silicon casts of individuals in his studio, then renders them into a pH-neutral, marine-grade cement that is twenty times stronger than its terrestrial counterpart. It is built to withstand the tremendous pressure of ocean currents and constructed of inert materials to encourage multispecies 'colonisation'. Because they are intended to become artificial reefs, the locations are shallow, chosen in consultation with marine biologists in order to both weaken the impact of hurricane-force currents on the shore, and to be strategically positioned for the 'seeding' of coral from one reef to another. As such, the sculptures are inordinately heavy, averaging one-to-two tons each, and are anchored to the seabed. Thus gravity and weight become constitutive elements in ensuring the futurity of multispecies sculptural ecologies.

These are permanent, 'swim through' exhibits, viewed while floating above and through the installation; unlike an interior, terrestrial gallery, the experience of the visitor is variable and dependent on skills and resources in snorkelling or scuba diving, as well as weather and currents; impressions are informed by light, viscosity of the water, age of the sculptures and presence of marine species. While the exhibits are 'permanent,' the sculptures are not; they change every day based on their occupation by bacteria and algae, and, eventually, coral. Environmental or earth art is known for its ephemerality, its locatedness, its participatory expectations and its pedagogical intents. It also reflects an entanglement with nonhuman force and a commitment to ecological regeneration.[9] While wind and rain may be the major elemental forces in transforming artworks on land, submarine aesthetics are subject to an alien environment, transformed by salt, currents, pressure and the rapid occupation by multispecies ecologies.

The submarine and material aspects of deCaires Taylor's eco-art are vital to its interpretation, particularly the ways in which water as a medium distorts time and alters knowledge production. After working for decades on the Law of the Sea, Borgese observed:

> the ocean is a medium different from the earth [...] it forces us to think differently. The medium itself, where everything flows and everything is interconnected, forces us to 'unfocus,' to shed our old concepts and paradigms, to 'refocus' on a new paradigm.
>
> (*The Oceanic Circle* 258)

This sense of newness, a critical engagement with an extraterrestrial space, raises questions about disciplinarity, epistemology, and (sea) ontologies. Critical ocean studies reflects an interdisciplinary approach to theorising the largest space on earth, but only recently has there been a discourse about how submersion may produce alternative knowledges and ontologies. Having worked as a diving instructor at the Great Barrier Reef before turning to underwater sculpture, deCaires Taylor observes:

> The experience of being underwater is vastly different from that of being on land. Objects appear twenty five percent larger underwater [and closer]. Colours alter as light is absorbed and reflected at different rates, with the depth of the water affecting this further. The light source in water is from the surface, this produces kaleidoscopic effects governed by water movement, currents and turbulence. Water is a malleable medium in which to travel enabling the viewer to become active in their engagement with the work.
>
> (quoted in Patel n. pag.)

Visitors to the sites describe their experiences of the sculptures as 'creepy,' 'spooky' and awesome, a sense of the shock of seeing specific human forms in the depths of a space deemed outside of human experience and temporality (quoted in Patel n. pag.). Casting all of the sculptures in terms of human scale adds a sense of both their diminutive size in relation to the vast ocean, as well as their particularly. It's this uncanniness, the engagements with human forms that appear distorted, larger, closer, and essentially 'matter out of place' to borrow from Mary Douglas (*Purity and Danger* 36), that lead to a phenomenological sense of their *unheimlich*, a feeling of that which is both familiar and radically defamiliarising. Thus a large man on a couch watching television would be an ordinary figure on land – his presence thirty metres under the sea is uncanny.

In Spivak's theorisation of 'planetarity' she seeks a different method of thought, like deCaires Taylor, in theorising alternatives to the homogenising and technocratic reaches of globalisation. Building on Levinas, she argues, 'To be human is to be intended toward the other. We provide for ourselves transcendental figurations of what we think is the origin of this animating gift: mother, nation, god, nature' (*Death of a Discipline* 73). To that list of the transcendental *unheimlich* I would add the ocean, a figure of evolutionary origin, amniotic sac and planetary future, suggesting a rupture in time as well as the unhomeliness of our experience of the planet. The ocean, like the 'figure of woman-as-mother-as-vagina' (*Death of a Discipline* 74) that Spivak reads in Freud's theory of the *unheimlich*, is also a figure for origins, an uncanny originary home. Of the vagina Freud writes, 'This *unheimlich* place [...] is the entrance to the former *Heim* [home] of all human beings, to the place where each one of us lived once upon a time and in the beginning' (quoted in *Death of a Discipline* 74). In a similar vein, deCaires Taylor has explained,

> For many years I've had incredible dreams about being underwater. My first instructor once told me [of these underwater caves] that it felt like being back in his mother's womb. I can relate. You have a much more detached consciousness underwater – like a form of meditation.
>
> (quoted in Patel n. pag.)

Thus the ocean is often figured as 'a species' of planetarity whether understood as the origins of earthly life, *mer* or mother, and more recently as a living 'super organism,' possibly supporting Lovelock's Gaia hypothesis of the earth as system.

These questions about temporality, place, multispecies life, and sea ontologies are brought together in one of deCaires Taylor's first sculptures, 'Vicissitudes' (2007), a circular structure of twenty-six children holding hands which invokes a cyclical model of time that will be transformed through multispecies occupations.[10] From the Latin vicis, to turn or change, 'Vicissitudes' reflects the engagement with an uncanny oceanic temporality and mutability. While the process of locating the sites, casting human forms, and submerging them into the sea takes months, deCaires Taylor points out that as soon as the sculptures are placed in the sea, time speeds up as they are rapidly covered with algae, one of the first steps in building a coral reef.[11] He has remarked that 'The sculpture proposes growth, change, and natural transformation. It shows how time and environment impact on and shape the physical body'.[12]

We see a ring of children facing outwards, holding hands, as they unite to face some kind of external force or event. As life casts, the details of the children's contemporary clothing are precise, as are their facial features. Standing barefoot in the white sand, surrounded by blue waters, the boys and girls seem both at home in this environment, yet simultaneously 'creepy,' as if responding to some threat that the viewer has not realised. In that way they are outside of our viewing time, responding to something that positions our own temporality as belated. On closer look, their eyes are closed, suggesting some kind of group concentration and communication, positioning the viewer as someone interrupting an intimate moment or prayer. As they are generally viewed from slightly above, the ring shape is the predominant figure; approaching the statues one comes to realise that they are positioned alternately boy and girl; a closer look reveals that they are the same two children, an uncanny doubling.

As soon as the photos of 'Vicissitudes' were released, people immediately interpreted it as a monument to the lost lives of transatlantic slavery. In the often-quoted words of Édouard Glissant:

> Whenever a fleet of ships gave chase to slave ships, it was easiest just to lighten the boat by throwing cargo overboard, weighing it down with balls and chains [...] These underwater signposts mark the course [across the Atlantic]. Navigating the green splendor of the sea [...] still brings to mind, coming to light like seaweed, these lowest depths, these deeps, with their punctuation of scarcely corroded balls and chains [...] The abyss is a tautology: the entire ocean, the entire sea gently collapsing in the end into the pleasures of sand, makes one vast beginning, but a beginning whose time is marked by these balls and chains gone green.
>
> (*Poetics of Relation*, 226)

An obsolete definition of 'vicissitudes' is reciprocation and return. Given the circular structure of the sculpture, mirrored by the ring of reinforced cement behind the children's hands that some have taken to be manacles, online speculation about its memorialisation to slaves led the artist to deny any intentional connection to the middle passage, while later acknowledging that in working with the tourist board he was forced to make compromises (in Patel n. pag.).

The Caribbean Sea has long been understood as a cultural and material space, a graveyard for the ancestors, a wound, an abyss or rupture in cultural continuity. While this shared experience, Glissant suggests, created an

> original victim floating toward the sea's abysses, an exception, it became something shared [...] Relation is not made up of things that are foreign but of shared knowledge. This experience of the abyss can now be said to be the best element of exchange.
>
> (*Poetics of Relation* 226)

It's not merely the peopling of the sea here that invokes the losses of the middle passage, but rather the 'shared knowledge' symbolised by the ring of children that suggests a submarine experience that cannot be shared by the terrestrial visitor, an experience of an ethereal otherworld of history, continuous with our own and yet simultaneously discontinuous. A submarine *unheimlich*, reflecting the mutually constitutive relationship between the ocean and the biography of the peoples of this place; foregrounding the ocean as uncanny medium which distorts our terrestrial-bound understanding of figures, time and space.

I'd like to pick up here on Glissant's gendered theory of the oceanic abyss as a 'womb abyss' (*Poetics of Relation* 226), which brings us back to the larger framework of planetarity and the *unheimlich* that I am arguing constitutes this oceanic imaginary. Glissant argues that the traumatic experience of crossing the abyss produced a particular kind of 'knowledge of the Whole, greater from having been at the abyss and freeing knowledge of Relation within the Whole' (226). He proposes a poetics of Relation that is represented in the 'aesthetics of rupture and connection [...] of a variable continuum, an invariant discontinuum,' constituted by interdependencies and entanglement (226). A transoceanic originary is vital to the recognition of the poetics of Relation: 'the unconscious memory of the abyss served as the alluvium for these metamorphoses' (226). Alluvium is sediment, a sand or soil from the Latin root 'to wash against,' invoking the erosion and transformation of matter, a vital part of deCaires Taylor's aesthetics.

Although Glissant's work has been engaged in terms of Caribbean and postcolonial ecocriticism, it has not been theorised in terms of what Deborah Bird Rose calls 'embodied knots of multispecies time' ('Multispecies Knots' 132). When asked about the visceral response to his submarine sculptures, deCaires Taylor replied it has

> something to do with the history of time and how we place ourselves in the evolutionary scale. [...] Underwater you are dealing with a completely different notion of time and it's confusing but fascinating; even for me when I return after a month and the work looks completely different.
>
> (quoted in Patel n. pag.)

As artificial reefs, these sculptures embody 'growth, change, and natural transformation,' vicissitudes, and that process is largely dependent on multispecies collaborations of the reef.

Relation: Reef ecologies

One of the indicators for the Anthropocene is the crisis in coral reefs across the global tropics. Largely understood as figures for biodiversity, 'rainforests under the sea,' coral reefs occupy less than 1 per cent of our oceans, yet are home to nearly one-third of known marine species. Coral is a eukaryotic multicellular animal. The reef is an ecology – a *zoophytalite* (animal-plant-mineral) form that grows so large that some can be seen from space. The hard coral of the tropics consists of a polyp surrounded by a limestone 'cup' that it secretes, like the shell of a turtle. These *zooxanthellae* reside in the tissue of coral, help it secrete calcium carbonate and remove waste. Like humans, coral are multispecies creatures; besides the symbiotic *zooxanthellae* there is a bacterial community in coral mucus that surrounds and protects it, not to mention the fish, turtles, hydrozoans, sponges, sea urchins, star fish and algae that are essential to its survival.

In addition to its figuring as a multispecies assemblage, coral signifies deep time. Its limestone structure means it layers growth like bone. As a colony, tube worms, bacteria and mollusks help build the infrastructure. It grows extremely slowly, some merely one centimetre a year, and its life span is undetermined. Some of the oldest (shallow water) tropical corals known are 500 years old, while some deep sea black coral is thought to be at least 4,000 years old. Recently scientists have speculated that the corals of the deep sea may be the ocean's oldest living organisms; due to their skeleton-secreting habits they form a living continuity with their ancestors and multispecies community.[13]

Ocean warming and acidification are causing a crisis in the symbiotic relations of a coral colony; the breakdown of these relations cause coral bleaching. It is said that over 50 per cent of Caribbean coral has been destroyed since the 1970s, due to ocean temperatures and acidification, sewage and agricultural runoff, overfishing and tourism. Reefs signal more than planetary crisis: 'coral are something to be read — for climate change, for potentially patentable genes, for representativeness' (Helmreich 'How Like a Reef' n. pag.).[14] Helmreich has explained that early western naturalists initially found coral to be 'an assemblage of flesh and stone that generated speculation about the boundaries of the living and nonliving' ('How Like a Reef' n. pag.). While the story of coral reefs has been anthropomorphised, like the concept of the Anthropocene, it has not been told with enough attention to the differentiation *between* the humans who contribute to environmental pollution. Threats to coral reefs are not evenly distributed – predatory practices of transnational corporations and neoliberal regimes put far more pressure on resource extraction from postcolonial states of the tropics than the Global North. Thus 'reefs are not just climatic barometers but also serve as indicators of North–South inequality' ('How Like a Reef' n. pag.)

This entanglement, as Glissant would have it, of the history of a submarine ecology like a coral reef with the human is brought to the foreground in deCaires Taylor's 'Vicissitudes,' where the two children whose features have been so precisely captured in sculpture are continually transformed into a more-than-human assemblage. In looking over the changes in the boy's face over the past few years we can see that the very human component that renders the initial sculpture to speak to 'sea ontologies,' the specificity of a 'shared knowledge' of Relation between the children, has been reconfigured. Their faces in particular, what function as the synecdoche for the human, are no longer recognisable; this visibility of the face ordinarily figures the human as a singular species, while creatures inhabiting the face, eyes, mouth or other facial features often signify a terrifying multispecies being akin to the futurity of science fiction. Now that his face is covered with algae, encrusting sponges and hydrozoans, the boy seems less accessible to human viewers, an alien, although all the more accessible to the multispecies bacteria and other creatures of the sea that establish coral reef ecologies. In her work on 'embodied knots of multispecies time,' Rose speaks of a 'generational time' that necessarily encapsulates death; a death that enables the existence of a new generation. Thus generational time is not necessarily understood in terms of species progressions, but multispecies sequences. In the transformation of the children of 'Vicissitudes,' we see this generational death, and to draw from Rose, 'we discern not [the] 'face,' but 'interface' (132).

There are (at least) three co-existing temporal strands that make up this 'knot' in deCaires Taylor's submarine sculpture. One might read the 'Vicissitudes' in terms of the living past: in the present participle, the middle passage lives at the bottom of the sea reflect 'hauntological time consciousness' (Baucom 31), a 'creepy' submarine visibility that signifies the historical abyss and 'these balls and chains gone green'. We might also read the transformed children as a future species, *aqua homo* of a multispecies Anthropocene, having merged with the other residents of an increasingly oceanic and tropical planet. Finally, we might read this in terms of the present, of the representation of two children from

Grenada facing an unspoken threat and a commentary on the contemporary environmental crisis in the Caribbean – one that is not anticipated (or even seen from outside the region) but currently experienced, and the necessity for multispecies alliances in navigating through this crisis in the present.

In 'Vicissitudes,' thinking with, interacting with, viewing, touching and perhaps being touched with what Eva Hayward terms 'fingeryeyes' opens up new opportunities for multispecies, submarine ontologies. Yet that should not take us outside of the local context to read a universal planetary crisis alone. This submersion brings about ethical transformations. As Nigel Clark has argued 'To be vulnerable to otherness, theorists of embodiment insist, is not just to be open to being unmade, but to being remade into something other than what we are [...] to being propelled in new and unforeseeable directions' (*Inhuman Nature* 249). These transformations, for the children of 'Vicissitudes,' are truly unforeseeable, as their very eyes have been reconfigured by algae into other kinds of sensing mechanisms.

The transformations required by these new sea ontologies are not to be confused with the type of multicultural liberalism championed by globalisation discourse; planetarity is the recognition of uncomfortable mergers, violence, digestion and indigestion (Haraway 'Anthropocene, Capitalocene, Chthulucene'), radical displacement from places, figurative and otherwise, that we may call home. I believe this requires a rethinking of the Anthropocene itself as the only model for imagining our futures, which has been readily critiqued for its anthropocentric bias.[15] As Anna Lowenhaupt Tsing asks:

> We become who we are through multispecies aggregations [...] This makes an enormous difference for our theories of 'human' action in the world. How can humans act as an autonomous force if our 'we' includes other species that make us who we are? [...] What might it mean for a multispecies aggregate to act upon the world?
>
> ('Strathern beyond the Human' 230)

The question as to who we mean by 'we' (what nations, classes, races, genders, ages, nations, regions, species?) when we speak of 'our' ecological futures is one of the great challenges to the Anthropocene. In a recent talk, Donna Haraway suggested that we put into play (at least) three different terms for figuring planetary futures. The Anthropocene, which helps to identify the human as (multi)species, the Capitalocene, which helps to locate the era in political, colonial and economic time, and a final term that I align with Spivak's concept of planetarity: 'the Chthulucene'. This term, Haraway argues, is for the 'chthonic ones, the not yet finished, ongoing abyssal and dreadful ones that are generative and destructive'. These are the subaquatic otherworldly others, because 'it matters to destabilize worlds of thinking with other worlds of thinking, it matters to be less parochial. If ever there was a time to need to be worldly, it is surely now. And I think all of us lack many of the skills' (Haraway 'Anthropocene, Capitalocene, Chthulucene' n. pag.).

Figure 16.1 Image and sculpture by Jason deCaires Taylor, 'Vicissitudes', Grenada, West Indies

I conclude with this last image of the children (Figure 16.1), one of whom looks towards us, perhaps expectantly, but as a 'moment in passing,' it's not clear from which world.

Notes

1 These developments are discussed in DeLoughrey, *Routes*.
2 There was ample black Atlantic scholarship before Gilroy; his work brought forward the oceanic contours, even if the ocean for him was not a material place. For the Pacific and Indian Ocean contexts, see Epeli Hau`ofa's *We are the Ocean* and Gaurav Desai's *Commerce with the Universe*.
3 See for instance DeLoughrey, Gosson and Handley's edited collection *Caribbean Literature and the Environment*
4 See www.seasteading.org/
5 I've adapted 'geontologies' for another suboceanic, multispecies focused paper, 'Ordinary Futures'.
6 www.underwatersculpture.com/sculptures/banker/
7 www.underwatersculpture.com/sculptures/inertia/
8 www.underwatersculpture.com/about/overview/
9 http://greenmuseum.org/what_is_ea.php
10 www.underwatersculpture.com/sculptures/viccisitudes/
11 www.youtube.com/watch?x-yt-ts=1421914688&v=-xR_uqkaflk&x-yt-cl=84503534#t=16
12 www.askmen.com/fine_living/travel/moilinere-bay-sculpture-park.html
13 www.llnl.gov/news/deep-sea-corals-may-be-oldest-living-marine-organism

14 Iain McCalman's recent book *The Reef: A Passionate History* also poses the interrelations between human and coral history, turning to how the reef has figured as a space of terror, nature and wonder in the Pacific.

15 DeLoughrey, 'Ordinary Futures'.

Works cited

2012. Dir. Roland Emmerich. Sony Pictures Home Entertainment, 2009. DVD.

Baucom, Ian. *Specters of the Atlantic: Finance Capital, Slavery, and the Philosophy of History*. Durham: Duke University Press, 2005. 31. Print.

Benítez-Rojo. Antonio. *The Repeating Island: The Caribbean and the Postmodern Perspective*. Durham: Duke University Press, 1992. Print.

Borgese, Elisabeth Mann. *The Oceanic Circle*. Tokyo: United Nations University Press, 1998. Print.

Chakrabarty, Dipesh. 'The Climate of History: Four Theses'. *Critical Inquiry* 35 (2009): 197–222. Print.

Clark, Nigel. *Inhuman Nature: Sociable Life on a Dynamic Planet*. London: Sage Publications, 2010. 249. Print.

Connery, Christopher L. 'The Oceanic Feeling and the Regional Imaginary', *Global/Local: Cultural Production and the Transnational Imaginary*. Eds Wimal Dissanayake and Rob Wilson. Durham, NC: Duke University Press, 1996. 284–311. Print.

——'*There Was No More Sea*: The Supersession of the Ocean, from the Bible to Cyberspace', *Journal of Historical Geography* 32.3 (2006): 494–511. Web.

Cousteau, Jacques. 'The Perils and Potentials of a Watery Planet', *Oceans: Our Continuing Frontier*. Eds H. William Menard and Jane L. Scheiber. Del Mar, CA: Publishers Inc. (1976): 13–17. Print.

deCaires Taylor, Jason. *The Underwater Museum: The Submerged Sculptures of Jason deCaires Taylor*. San Francisco: Chronicle Books, 2014. Print.

'Deep Sea Corals May Be Oldest Living Marine Organism'. *Lawrence Livermore National Laboratory*. 2009. Web.

DeLoughrey, Elizabeth. 'Ordinary Futures: Interspecies Worldings in the Anthropocene', *Global Ecologies and the Environmental Humanities; Postcolonial Approaches*. Eds Elizabeth DeLoughrey, Jill Didur, and Anthony Carrigan. New York and London: Routledge, 2015. 352–372. Print.

——*Routes and Roots: Navigating Caribbean and Pacific Island Literatures*. Honolulu: University of Hawai'i, 2007. Print.

DeLoughrey, Elizabeth M., Renée K. Gosson and George B. Handley. *Caribbean Literature and the Environment: Between Nature and Culture*. Charlottesville: University of Virginia, 2005. Print.

DeLoughrey, Elizabeth, Jill Didur and Anthony Carrigan. *Global Ecologies and the Environmental Humanities: Postcolonial Approaches*. New York and London: Routledge, 2015. Print.

Desai, Gaurav Gajanan. *Commerce with the Universe: Africa, India, and the Afrasian Imagination*. New York: Columbia University Press, 2013. Print.

Douglas, Mary. *Purity and Danger: An Analysis of the Concepts of Pollution and Taboo*. London and New York: Routledge, 1996. Print.

Earle, Sylvia A. *Sea Change: A Message of the Oceans*. New York: G.P. Putnam's Sons, 1995. Print.

Fagan, Brian M. *The Attacking Ocean: The Past, Present, and Future of Rising Sea Levels*. New York: Bloomsbury, 2013. Print.

Giddens, Anthony. *The Consequences of Modernity*. Stanford, CA: Stanford University Press, 1991. Print.

Gilroy, Paul. *The Black Atlantic: Modernity and Double Consciousness*. Cambridge, MA: Harvard University Press, 1993. Print.

Glissant, Édouard. *Poetics of Relation*. Trans. Betsy Wing. Ann Arbor: University of Michigan, 1997. Print.

Hayward, Eva. 'Fingeryeyes: Impressions of Cup Corals', *Cultural Anthropology* 25.4 (2012): 577–599. Print.

Haraway, Donna. 'Anthropocene, Capitalocene, Chthulucene: Staying with the Trouble', Anthropocene Conference, University of California, Santa Cruz. 9 April 2016. Web.

Harvey, David. *Justice, Nature, and the Geography of Difference*. Oxford: Blackwell, 1996. Print.

Hau'ofa, Epeli. *We Are the Ocean: Selected Works*. Honolulu: University of Hawai'i Press, 2008. Print.

Helmreich, Stefan. *Alien Ocean: Anthropological Voyages in Microbial Seas*. Berkeley: University of California, 2009. Print.

——'How Like a Reef: Figuring Coral, 1839–2010', *Essays in the Anthropology of Biology and Beyond Sounding the Limits of Life*, 2010. Web.

McCalman, Iain. *The Reef: A Passionate History*. Sydney: MacMillan Publishers, 2013. Print.

Mentz, Steven. 'Shipwreck Modernity'. Blog post. 14 February 2016. Web.

Patel, Amar. 'From the Vaults – in Conversation with Underwater Sculptor Jason Taylor', *RSS*. 3 November 2014. Web.

Povinelli, Elizabeth A. 'Geontologies: a Requiem to Late Liberalism' (2013). Paper presented at The Anthropocene Project, Hans der Kulturen der Welt, Berlin, 10 January. Web.

Pratt, Mary Louise. 'Planetary Longings: Sitting in Light of the Great Solar TV', *World Writing: Poetics, Ethics, Globalization*. Ed. Mary Gallagher. Toronto: University of Toronto Press, 2008. 207–223. Print.

Rediker, Marcus. *Between the Devil and the Deep Blue Sea: Merchant Seamen, Pirates, and the Anglo-American Maritime World, 1700–1750*. Cambridge: Cambridge University Press, 1987. Print.

Rising Waters. Dir. Andrea Torrice. TorriceMedia, 2002. Documentary series.

Rose, Deborah Bird. 'Multispecies Knots of Ethical Time', *Environmental Philosophy* 9.1 (2012): 127–140. Print.

Spivak, Gayatri Chakravorty. *Death of a Discipline*. New York: Columbia University Press, 2003. Print.

Steinberg, Philip E. *The Social Construction of the Ocean*. Cambridge: Cambridge University Press, 2001. Print.

Steinberg, Philip E. 'Of other seas', *Atlantic Studies* 10.2 (2013): 156–69.

Steinberg, Philip E., Elizabeth Nyam and Mauro J. Caraccioli. 'Atlas Swam: Freedom, Capital, and Floating Sovereignties in the Seasteading Vision', *Antipode* 44.4 (2011): 1532–1550. Print.

Stoler, Ann Laura. 'Imperial Debris: Reflections on Ruins and Ruination', *Cultural Anthropology* 23.2 (2008): 191–219. Print.

Time and Tide. Dir. Julie Bayer and Josh Salzman. CreateSpace, 2005. Documentary.

Tsing, Anna Lowenhaupt. 'Strathern beyond the Human: Testimony of a Spore', *Theory, Culture & Society* 31.2–3 (2014): 221–241. Print.

Walcott, Derek. *Collected Poems, 1948–1984*. New York: Noonday, 1988. Print.

'What Is Environmental Art?' *What is Environmental Art?* 10 April 2016. Web.

Zalasiewicz, Jan (2008). *The Earth After Us: The Legacy That Humans Will Leave In The Rocks*. Oxford: Oxford University Press. Print.

Index

Index entries with 'n' denote note numbers.

Printed in the United States
By Bookmasters